The In-House Option: Professional Issues of Library Automation

CONTRIBUTORS

T. D. Webb is Assistant Professor of Library Science, and Coordinator of Library Automation at the Joseph F. Smith Library, Brigham Young University-Hawaii Campus. He was formerly on the staff of Phoenix Public Library, and worked in the Technical, Public, and Extension Services divisions.

D. Errol Miller is Assistant Professor of Educational Media, and Supervisor of Public Services at the Joseph F. Smith Library, Brigham Young University-Hawaii Campus. He is also supervisor of the Smith Library's Media Production Laboratory.

The In-House Option: Professional Issues of Library Automation

T. D. Webb

Routledge
Taylor & Francis Group
NEW YORK LONDON

The In-House Option: Professional Issues of Library Automation is #1 in the Haworth Library and Information Sciences Text series.

First published 1987 by The Haworth Press, Inc.

Published 2020 by Routledge
605 Third Avenue, New York, NY 10017
2 Park Square, Milton Park, Abingdon, Oxon OX14 4RN

Routledge is an imprint of the Taylor & Francis Group, an informa business

ISBN 13: 978-0-86656-617-9 (hbk)

Library of Congress Cataloging-in-Publication Data

Webb, Terry.
    The in-house option.

    (The Haworth library and information sciences text; #1)
    Bibliography: p.
    Includes index.
    1. Libraries—Automation. 2. Libraries—Automation—Case studies. I. Title. II.
Series.
Z678.9.W4   1987   025'.0028'54   87-127
ISBN 0-86656-617-1

CONTENTS

Preface

Librarianship is currently facing a number of professional issues that not only reflect the dramatic changes occurring in library science, but also portend the future of information services and of the people and institutions that will be providing them. Many of these issues arise from the automation of libraries. Indeed, the situation is very much as Freeman fears:

> Today there are many observers who see fissures within the profession growing wider in the future. One writer worries that the great unifier, the computer, may actually lead to further divergence. Unlike the 1960s and 1970s, when financial necessity forced thousands of librarians to band together in support of needed bibliographic utilities, the 1980s with its microcomputers may lead each individual library to innovate on its own and not in combination with others. It is also possible that libraries will be distinguished by their ability or inability to afford the costs associated with advancing technology. This makes the need for theory even more pressing.[1]

Modernization of any kind, be it within a primitive tribe or an already technologically sophisticated profession, brings with it challenges that create new issues and concerns, and lays added stress on any questionable areas of the prevailing structures of belief and action. The result, in either case, is a rearrangement that is a synthesis of the old and the new. Although there will be a high degree of continuity between the old and new systems, the new entity can be expected to have an adaptive advantage over the old, having acquired traits that better accommodate the environmental change.

This certainly seems to be the case with library automation. No other event compares with its modernizing effect on library services. No other topic has generated the same output in library literature or upheaval in the professional duties of librarians. The computer promises to renew the profession of librarianship just as it has so many other professions. Yet as Freeman suggests, the theoretical basis for the synthesis remains obscure. The issues and professional implications around which the movement to renew the profession will form remain largely unarticulated.

In short, it is time to assess the impact of automation on professional librarianship in terms of individual practitioners, institutions, and underlying philosophy. The present study cannot hope to supply the type of exhaustive theory that Freeman finds so lacking and that he believes will unify the profession in the midst of change. But perhaps it can at least identify and document some of the professional issues of library automation.

My experience indicates that some of the major issues, computer-related and otherwise, facing the library profession at present are

1. the level of computer literacy among professional librarians;
2. modal personality traits and educational backgrounds among professional librarians;
3. appropriateness of library education to the changing demands of the profession;
4. methods of attracting candidates with the needed interests and skills to the profession;
5. compatibility of automation and the philosophy of librarianship;
6. the role of professional librarians in automated systems design;
7. the role of the library within the larger governmental-political-economic environment.

These issues are, of course, closely interrelated and are felt to one degree or another throughout a number of library operations. But in certain operational situations, they combine in such a way as to define the current state of librarianship in its process of modernization. From these significant instances can be drawn a set of strategies for directing the future course of the profession.

Rather than deal with the issues singly, or in a wholly abstract or unfocused way, I have chosen to ground the discussion in practical examples of the decision-making situation, as this seems to me to be the clearest point at which the issues converge. Furthermore, it is a situation that every library must face when it decides to purchase an automated system.

All the issues cited above converge most meaningfully and are most amenable to practical analysis in a consideration of host computer location options. The options are (1) locating the computer in-house to be operated by librarians who, though expert in bibliography, may have limited expertise in the handling of computers; and (2) installing the host computer instead at a remote computer center belonging to the library's parent organization where system maintenance will be the responsibility of personnel who are experienced in computer operation but know little or nothing about librarianship. The relative merits of each type of system are important to the decision making of libraries undergoing automation or contemplating a change of automated systems. The sophistication of the new systems has altered the nature of system configuration such that locating the host in house may not be as advantageous as may appear.

Yet library managers often make the choice with little or no awareness of the comparative advantages and disadvantages of the two configurations. Thus a comparison will not only clarify the merits of the two, but will underscore the professional issues of library automation involved. For while making the decision to locate in house or at a remote site, the library manager, perhaps unwittingly, confronts every major issue mentioned above. Articulating them within the actual automation context makes them vividly apparent.

Many libraries, of course, have no choice in the matter of host computer location. Some libraries must house the computer in the computer center on campus, for example, while others may be required to share time on a machine already installed "downtown" in a municipal automation department. Some smaller or otherwise independent libraries may not have access to a computer center at all. But many libraries do have a choice in the matter or can at least make their preferences heard. Libraries that have no options should know what problems they will encounter with their automation project whichever configuration they are dealing with. Libraries fortunate enough to have a say in the matter need a basis for a wise decision.

It is almost a certainty that in the future an increasing number of libraries will have an option with respect to host location because more and more academic and municipal computer centers are agreeable to the idea of computer resource "deployment," that is, the distribution of mainframe computers throughout the organization. This is because computer applications and availability are becoming too widespread to be managed effectively from any single computer center. Hardware deployment, however, does not usually involve the simultaneous deployment of the expertise needed to effectively operate the equipment in question.

This book will be based on two in-depth case studies. The first is from the Phoenix Public Library, which uses the ULISYS automated system installed on a host computer in the city's Management Information Systems (MIS) department. The second study is from the Joseph F. Smith Library at Brigham Young University-Hawaii Campus, which utilizes a Dynix integrated system based on an in-house minicomputer.

Little has been written on the topic of host location, and what there is deals almost exclusively with cost comparisons. I have found, however, that a library's decision on where to locate its computer is one of the most consequential decisions in an automation project. The ramifications of that decision affect not only the day-to-day effectiveness of the library but also future trends in librarianship with respect to the issues enumerated above.

Among the large number of otherwise excellent books that treat the subject of library automation, none discusses the topic of host computer location in sufficient depth. They do not address the unsuspected complexity of this decision and its consequences. Nor do they consider the larger professional issues entailed in the decision.

The present volume, then, is designed to be on one level a practical guide to host computer location, with examples of both remote and in-house configurations that show the strengths and

weaknesses inherent in both. But on another level the book artic-
ulates the rather broad professional issues of library automation.
It is hoped this will be beneficial to individual libraries involved
in automation planning and to the evolving profession of librarianship.

Thus the "in-house option" of the title has a double meaning.
It refers to the strategy of an in-house hardware configuration for
a library's automated system. It also has a deeper meaning that
refers to momentous issues in the internal structures of professional librarianship that need to be resolved. Either the profession must broaden itself to include those technical areas intro-
duced into libraries by the sudden and complete advent of
automation, or librarianship will be parceled among diverse
groups of specialists, of which librarians will be one among the
many.

REFERENCE

1. Michael Stuart Freeman, " 'The Simplicity of His Pragmatism': Librarians and
Research," *Library Journal* 110 (9) (May 15, 1985): 29.

Introduction

At the end of the last decade, Allen B. Veaner asked librarians this question: "What Hath Technology Wrought?"[1] He then looked at the state of library automation and the overall impact of the computer on libraries and library service up to that time and made some very perceptive evaluations in a number of key categories.

His generally negative appraisal centered largely on three weaknesses in the area of library automation. First, library systems, he said, were overly sophisticated. When the computer experts and systems designers, whom Veaner called "computerniks" (none of whom were trained librarians), asked librarians what we wanted in our systems, we answered, "We want everything!" And the result was rigid, expensive, complicated, nonfriendly systems that did everything, except make life simpler. Then we began to learn that the more sophistication we built into our systems, the less room there was for user-friendliness and the greater the need for highly sophisticated and costly personnel to maintain and further develop those systems.

Second, Veaner pointed out that in the early phases of library automation, there was an overall lack of direction from the library profession. As he said, "For a long time, the library profession permitted the technological tail to wag the bibliographic dog."[2] Admittedly, any system will be constrained by the available technology. Hence, the lack of user-friendliness that Veaner speaks of was partly a result of the technological limitations prevailing in those early days, which have now to a large degree been solved. Yet many features, drawbacks, and options that were inherent in the existing technology were not explored because librarians and non-librarian computerniks were not communicating well. Veaner attributed this to the over-ambitiousness of the computerniks and to a general lack of management skills and drive among librarians, who traditionally receive little or no professional training in administrative practices.

To this I would add that I have also observed in librarians a startling lack of mechanical aptitude. And as the following chapters will discuss, I believe this engenders in librarians a certain apprehensiveness toward the computer—which is, after all, only a machine—as well as all of its online and offline peripherals. As a result, instead of defining the specific output desired for their envisioned systems (which is a remarkably difficult task under any circumstances), librarians, out of a fear of machines, relinquished much of the vital responsibility for planning to the computer experts.

Veaner attributed these two problems of library automation to a third, which, he said, could be called a professional hubris among librarians, an "overbearing sense of self-importance" that caused them to miscalculate the public's esteem for bibliography. I have elsewhere addressed the general lack of reliable data to demonstrate the real dollar value of libraries to the public and the detrimental effect this has on library planning and financial stability.[3] But this lack of information and perspective has been especially apparent, according to Veaner, in library automation. It has led not only to overblown computerized systems, but also to a disregard for the costs of developing and maintaining these systems.

Despite his disapproving report, however, Veaner ended his summary with hope. The profession may yet right itself in the matter of library automation. He says,

> I would prefer to title some future review "What Have *We* Wrought?" in the hope that some day, we'll be wise enough to have exercised adequate professional leadership—which will not only assure our survival but also guarantee that we survive as the masters of technology, not as its slaves.[4]

The situation, even when Veaner wrote those words, was probably not as dismal as he reported it. But he made his point. The mistakes were real, and their consequences were inescapably longlasting. Yet there remains hope, for in Veaner's terms, "a mistake is not a tragedy."

This book is certainly in the vein of Veaner's hard look at library automation, but I believe the library profession is still not ready to write "What Have *We* Wrought?" Systems still tend to be overblown by some standards, and library administrators are not yet providing the degree of direction needed in the automation process. Likewise, there still remains a gap in mutual comprehension between librarians and Veaner's "computerniks."

What this book attempts is to reduce the slavery of libraries and librarians to technology. I hope to demonstrate that librarians are now in a better position to become the masters. At least, many more options are available now with respect to automation.

One of the options is the choice between basing the library system on a host computer owned and operated by the library's parent organization — city, university, school, etc. — and utilizing an in-house, stand-alone system based on a computer located on the library premises and operated by library personnel.

As the following case studies will show, both the Phoenix Public Library and the Joseph F. Smith Library at Brigham Young University-Hawaii Campus went through an early phase of automation that utilized a remote host computer, although their situations were quite different. In brief terms, Phoenix Public utilized a dedicated minicomputer housed and operated in the City's Management Information Systems (MIS) Department several miles away. The software was a vendor-maintained package similar to those available on the current market. The Smith Library shared time on a university-owned mini located across campus from the library in the Computer Services department. The software, however, was locally developed and maintained by the library staff and student-worker programmers from the university's computer studies department.

At this writing, the Smith Library has abandoned its locally developed software and remote host arrangement, and purchased an in-house mini and a turnkey integrated system. Likewise, Phoenix Public is designing the specifications and output requirements for a new system that may possibly be based on an in-house, stand-alone host computer.

Of course, there are other versions of the remote host arrangement. For instance, a mainframe at a centralized facility may provide services via telecommunications lines to many libraries, where the only in-house hardware consists of terminals, modems, and printers. A distributed system, on the other hand, decentralizes certain types of processing among members of a consortium. In this case, each participating library may have an in-house computer, but because no single unit provides all processing services for its library, each library must be served to one degree or another by a remote machine.

Every remote host arrangement places a unique set of demands on a library and requires its own style of managerial maneuvering to deal with problems. But remote host configurations are becoming less common as more and more libraries purchase turnkey systems from vendors who assume that the library manager will prefer to have the supporting hardware within the library facility. And library managers are proving only too willing to go along with this assumption. The question is; Does this decision reflect a carefully planned management strategy to assure the best service potential for the patrons? Or is it instead another instance of the lack of leadership and of the "overbearing sense of self-importance" Veaner spoke of?

A major concern of this study is that many library administrators may be leaping to the in-house option too quickly because a vendor tells them that they should and assumes they will, or because the manager reasons that it is better to locate in house simply because the technology to do so is available. The point of this book, then, is to give the librarian more grounds for decision making by reinforcing the fact that in-house location of system hardware is only an option, and should not be considered a matter-of-fact and incontrovertible condition of library automation.

This is not to say, of course, that a decision to install an in-house system is a faulty piece of library management, nor that having a remote host represents a wise one. The goals of a particular library, its size and the composition of its staff are among the factors an administrator must consider before making such a decision. What is right for one library may be dead wrong for another.

Host location is not always regarded as a crucial element in a library's automation plan. The important items are considered to be such things as price, disk storage capacity, response time, maintenance fees, software capabilities, etc. These are major concerns, of course, and library literature abounds in reports from authors giving advice about decision making in each of these important areas. My experience tells me, however, that other elements, including host location and operation, are also important. These same reports, although they warn that someone will have to be trained to care for the hardware in an in-house configuration, do not detail how involved the training must be in order to look after the complex and sophisticated machinery involved and to make certain that it is operated in the most efficient and effective manner possible.

The crucial factors here have to do with the operation of the computer and with the current level of computer expertise among librarians. I do not believe librarians as a rule are equipped to operate an in-house minicomputer effectively. It is not that librarians lack intelligence. Let it be admitted that we are a bright lot. But rather that computers are still difficult to use despite, or because of, all the recent sophistication that change has brought in them. Effective use of a large computer requires a great deal of training, much more, claims to the contrary notwithstanding, than any vendor is ready or willing to supply.

Simply speaking, computers are difficult to master. User-friendliness is a vendor-defined concept, one I have found to be largely a myth. While it may be true, as many authors have pointed out, that librarians will not have to know programming to operate the new minicomputers and the turnkey software based on them, the level of experience and knowledge required to operate all the components of an in-house system still exceeds the expertise generally available in libraries. Moreover, the number of librarians who possess such experience is increasing slowly, certainly at nowhere near the rate at which libraries are installing in-house computers.

Effective use of an in-house, stand-alone, mini-based system requires a familiarity with such things as modems, telecommunications lines, printers, query languages, report generators, terminals, terminal emulators, file structure, system architecture, data base construction, and much more — to say nothing of the actual operation of the central processing unit itself. Troubleshooting can be very sticky, and some vendors require some degree of local diagnostics before they will accept a call.

As a result of hidden complexities like these, a number of small businesses, yet larger than some libraries now involved in in-house arrangements, have reversed their original automation projects, sold their hardware, and contracted with computer firms to perform the projects needing automated assistance. In most cases, the automation requirements of these businesses involved payrolls, billing, accounting, and other similar needs that are much simpler to automate than bibliographic description, circulation control, and database searching.

In addition, the application of computer technology has recently crossed a new frontier, namely, that of decision support. This new application addresses the needs of administrators and professionals. Compared to the automation of the clerical sector, which has thus far been the primary concern of computer application in the business community, automation of the managerial sector promises to be very difficult because administrators and professionals are not governed by routines, policies, or repetitive tasks, but by the contingencies of the moment and the parameters of the problems to be solved.

Library managers will certainly want to take advantage of these applications as they become available, and clearly managers with in-house computers will think they have the resources at hand to do this. But these new fields of computer application will require an even greater amount of expertise and technical support, especially during their early stages of development and implementation. They will surely put an additional strain on any library automation department staff, particularly when there is limited expertise to begin with.

Thus the question becomes; Do we install the computer in-house and train librarians in the things they need to know in order to manage the system effectively? Or, do we locate the computer at an existing computer facility and train appropriate automation experts from that facility in what they need to know about librarianship?

The case studies included in this volume are intended to help library managers arrive at the best possible answer to these questions for their respective institutions. Phoenix Public chose the remote host option, hoping it could adequately communicate the library's needs to the professional computer operators tending the system at the city's MIS Department. The Smith Library, on the other hand, chose an in-house system. These two studies detail the problems any library choosing one or the other option can expect to face.

To summarize, then, this book will be a comparative analysis of some of the technical and administrative conditions inherent in systems utilizing either a remote host computer operated by professional programmers or an in-house computer located in the library and operated by library personnel. The purpose of the book is to provide information upon which library administrators and planners can base automation decisions affecting system design, implementation and utilization. It is hoped that this will take the library profession a step closer to Veaner's fabled day when librarians can look at their institutions and assess what they have done with technology, and not regret what automation has done to them as professionals and to the library as a learning institution.

It is significant that the new automation options available to library managers, of which remote host/in-house is only one, have come about largely through recent hardware and software improvements. Yet the failings of the library automation movement that Veaner cited and that stem from the earliest phases of that movement are not directly machine-related. Instead, they result primarily from human failure to interact well with a new technology. The same holds true for many of the problems that will be addressed here. To clarify some of these failings in order that they may be resolved is the intent of this study.

REFERENCES

1. Allen B. Veaner, "What Hath Technology Wrought?" in *Clinic on Library Applications of Data Processing, Proceedings, 1979: The Role of the Library in an Electronic Society*, ed. by F. Wilfrid Lancaster (Champaign, IL: University of Illinois, Graduate School of Library and Information Science, 1980), p. 3.

2. Veaner, p. 5.
3. T. D. Webb, *Reorganization in the Public Library* (Phoenix, AZ: Oryx Press, 1985), pp. xiv-xvi. See also Malcolm Getz, *Public Libraries: An Economic View* (Baltimore, MD: Johns Hopkins University Press, 1980), p. 172.
4. Veaner, p. 14.

Chapter 1

Library Automation:
The New Shamanism

Among organizations employing large numbers of professionals, libraries are marked by a commonality of knowledge shared at all professional and administrative levels. Regardless of the degree of horizontal differentiation or departmentation within a library, all the professionals share a common body of training and expertise. Of course, different librarians in different departments may perform widely differing tasks. Nevertheless, those tasks generally are integral features of librarianship, and all the professionals in the library have had at least some grounding in them all during their training. For instance, not all librarians are catalogers, but all librarians are to an extent familiar with classification schemes and the mechanics of bibliographic description. Furthermore, they all make regular use of this knowledge to one degree or another. In this sense, libraries generally typify, with some variance, the professional organization described by Hall:

> The organization of the professional departments tends to be rather "flat." That is, the concept of equality among professionals (lawyers should be given equal status within a legal department) leads to little differentiation within a department. The model followed is collegial, which is characterized by a few formal distinctions or norms and by little in the way of formalized vertical or horizontal differentiation.

He says further,

The work of the professional is likely to be highly complex. Within a research division, university department, law firm, or hospital, there can be extreme diversity in the activities being performed. There is an intensive division of labor. This complexity is not organizationally imposed, however. It is a consequence of the differentiation of the activities among the professionals and is not a formal organizational characteristic.[1]

In terms of administrative hierarchy, library managers almost invariably are themselves librarians and thus share in the common body of professional knowledge and training. In this regard, libraries resemble law firms in which the directors tend to be attorneys, but they differ from clinics and hospitals, in which administrators typically are not physicians. In professional organizations, as Etzioni says, such an arrangement can be counterproductive: "When people with strong professional orientations take over managerial roles, a conflict between the organizational goals and the professional orientation usually occurs."[2]

What is happening in libraries is a case in point. More and more the controlling knowledge of a particular non-professional area of modern library technology is being relegated to a select few who are non-librarians, while the technology itself is becoming increasingly pervasive throughout library functions and operations performed in fact by librarians. As a result, a disequilibrium of knowledge now exists in library practice that can radically affect libraries and librarians. As Etzioni says, "The basis of professional authority is knowledge, and the relationship between administrative and professional authority is largely affected by the amount and kind of knowledge the professional has."[3] Thus, as librarians lose sight of the knowledge that controls the library as an organization, so their institutional authority is diminished.

Of course, the technological area spoken of here is library automation, and the controlling knowledge, that is, the knowledge disequilibrium, is computer literacy. In short, the current state of library automation requires more expertise than the average librarian possesses at this point. The relatively few with the necessary expertise are, in fact, too few to support the growth rate of library automation. These are major premises of this book.

Analyzing the disequilibrium will indicate the nature and the breadth of the gap between the computer expertise of the average librarian and the level of computer competency necessary to operate effectively the type of computer system now available on the market. The analysis will require a careful look at librarians themselves, their backgrounds and common traits, and an evaluation of the current and projected state of library automation. The end result, it is hoped, will be a clearer picture for library managers of certain types of hidden "costs" of automation, with particular regard to the strategy of locating the computer in-house.

The first step, however, is by way of a discussion of the concept of computer literacy as it applies to libraries. It should be noted that "computer literacy," as it is used in the context of the present study, refers specifically to mini and mainframe computers and the integrated library systems they host, which are now being developed and marketed by turnkey vendors.

COMPUTER LITERACY

Computer literacy has become the subject of controversy. There are at least three schools of thought as to what is actually meant by computer literacy. To the first school of thought, it means simply a familiarity with a basic automation vocabulary and an awareness of areas suitable for computer applications.[4] This view will not be discussed in the present study. Another view, common among computer scientists requires computer users to be proficient programmers in multiple languages and to be thoroughly familiar with machine operation.[5] Anything less does not qualify as literacy, and all computer involvement, according to this view, assumes this level of user expertise. A third view, one gaining favor in professions outside computer science, emphasizes computer skills as important, but ancillary to basic professional problem-solving ability.[6] In other words, following this reasoning, businesses do not want computer scientists with a minor in business administration, they want MBAs who know their way around computers.

There are other views and variations, but these last two represent the major approaches taken by computer educators on the subject of computer literacy. A report from the Department of Defense Dependent Schools (DoDDS) system casts the two approaches and their respective competencies in terms of instructional objectives (see Appendix A), and distinguishes them as "computer science" and "computer literacy."[7] These convenient labels will be adopted for the remainder of this study.

In essence, the "computer science" approach requires a person to know more about computers than may actually be needed outside the field of computer science itself, while the other view requires him or her to know only what is needed for use within specific professional disciplines. It must be remembered, of course, that the literacy requirements are different for different professions. Part of the purpose of this study is to discuss what the requirements are for librarianship, although the degree of variation between individual libraries may be as great as that between different professions. The case studies in Chapters 4 and 5 illustrate the level of literacy required by two specific libraries. But each library manager must evaluate the literacy requirements demanded by the library's automation plans, if those plans are to be optimally successful. In a larger sense, there is also a basal literacy level for the library profession at large, which should be considered and which has yet to be measured.

LIBRARIANS AND COMPUTER LITERACY

Issues comparable to these are being addressed in other professions, such as medicine, business, and education. These professions, albeit somewhat after the fact, are attempting to strike a balance between computer savvy and professional expertise based on the needs of a profession and those served by it and, perhaps to a lesser degree, on the characteristics of the professionals in the field.

Librarianship needs to follow suit. We have yet to find our proper automation balance. The effects of the imbalance are abundantly apparent in the unhappiness of many librarians over what their systems will and will not do for them, and over what sorts of keyboard and logical contortions are necessary to make the systems deliver. The argument of this study is that we approach this imbalance of system performance and professional literacy by looking first for the basal literacy level of professional librarians in the field, and then gear automation somewhat above that level. One must assume that the basal level is rising at a significant rate. Yet many automation projects seem to overshoot that level entirely. Failure to proceed in this manner, that is, to continue to ignore the literacy level while producing systems of ever-increasing sophistication, risks alienating librarians from their own profession.

One possible way to estimate the basal computer literacy level among practicing librarians is to administer a test like the Computer Literacy Examination: Cognitive Aspect (CLECA) recently developed by Cheng and Stevens (see Appendix B). The test is designed to provide a standardized method of measuring the achievement levels of high school students (11th and 12th grades) in computer literacy courses that are primarily for "students who will not become computer professionals."[8] Its authors write,

> The CLECA was designed to focus on two major themes: (1) awareness about computers, and (2) basic programming skills. Seven cognitive-related topics were used as the basis for test construction. These topics include: computer terminology, computer language commands, writing computer programs, parts of computers, writing algorithms, math concepts, and history of computers.[9]

Admittedly the CLECA is designed to measure literacy progress, not basal competency, and its target group is high school students, not professionals in the information field. Furthermore, it focuses on a specific brand of microcomputer, not a minicomputer or mainframe of the type to be dealt with in this study. Nevertheless, the CLECA corresponds reasonably well to the computer literacy portion of the DoDDS instructional objectives. Also, as will be seen later, the CLECA measures specific types of competencies that are commonly required in library automation projects, including the new turnkey systems. Such a measurement, therefore, not only provides empirical data concerning the basal literacy level, but also correlates the integrated systems being developed to the current library marketplace.

In the absence of such precise data drawn specifically from the library profession, however, other methods must be used to estimate the growing disequilibrium of knowledge that exists between the basal computer literacy of librarians and that required to operate a powerful computer effectively. One method is to formulate an "impressionistic" analysis of the behavior of librarians in an automated context. The second is to apply to librarians the findings of appropriate studies conducted in other closely related disciplines. The impressionistic analysis, of course, measures nothing. But it may accurately describe the attitudes that underlie and therefore bring about the prevailing condition of computer literacy among librarians, whatever that condition may be. From there, we may abstract the condition and predict its course. The cross-disciplinary borrowing cannot safely be said to represent actual conditions in libraries, but at least it may indicate a probable level of literacy.

With respect to the first method, there seems to exist among librarians a psychological state some call "the terminal barrier." In his study of patron use and non-use of online catalogs, Kaske states that

> there are at least three barriers to the use of online catalogs. These barriers are:
>
> 1. computers
> 2. search mechanics of a system
> 3. bibliographic data
>
> In order for people to use online catalogs today, they must first dispel their fear of computers and learn to use a computer terminal. Once they have a basic knowledge of computer terminals and their operation, the second barrier—the search mechanics of a system—must be overcome. At this time, all systems use different search me-

chanics. The final barrier is that of the bibliographic data itself.[10]

Of course, Kaske is talking about patrons. Librarians are not likely to fear bibliographic data. Yet Kaske, like most others who write on this topic, fails to address the possibility that the other two barriers — fear of computers and the system architecture — could be prevalent among librarians and patrons alike. Indeed, I have witnessed five different behavioral manifestations of this condition. First, the librarian fears that an improper input may somehow harm the computer, and is, therefore, reluctant to interact with it. Second, the librarian is unable to supply the proper commands at the appropriate prompts despite repeated training sessions. Third, the librarian ignores the computer completely and relies on more traditional methods to serve the patron. Fourth, the librarian overreacts to the machine challenge and becomes obsessed with it to the near exclusion of other professional concerns, making work with the computer almost a fetish. Finally, the librarian is afraid to input data because someone later will be able to view any mistakes made, thus creating a type of online stage fright.

Such behaviors, although they sound ludicrous, are evidence of a real psychological impediment to dealing with an interactive machine. Among librarians, these behaviors are sometimes referred to as "the fear of inputting."[11] Elsewhere they are given labels such as technophobia or cyberphobia. Brod calls the condition "technostress," and describes yet another behavioral abnormality — "flaming" — which is "making rude or obscene outbursts by computer."[12] Although I have not observed any flaming among librarians, a few celebrated cases have appeared in the literature.[13] In simple terms, these conditions are the results of anxieties arising from the suspicion that computers, in the words of one librarian, "are certainly unforgiving if not downright sinister."

Eventually librarians will learn to control online catalogs. But in so doing will they actually overcome their fear of the machine? If not, how expert and reliable can their manipulation of a powerful computer really be? Furthermore, will the fear of inputting be more prevalent at the deeper levels of interaction where true computer literacy and even computer science are necessities? The fear of inputting exists at all levels of computer interaction. Among the newly initiated it may manifest itself as a stupor of thought or near-paralysis whenever the computer produces an unexpected response on the display screen. Among the ranks of the "experts," the fear is sometimes manifested as a mimicking of the intricacies of computer logic and a ritualistic use of computer jargon. Brod says that technocentered people "begin to adopt a mindset that mirrors the computer itself":

> The primary symptom among those who have too successfully identified with the computer technology — those for whom the computer has become the central core of existence — is a loss of the capacity to feel and to relate to others. . . . Signs of the technocentered state include a high degree of factual thinking, poor access to feelings, an insistence on efficiency and speed, a lack of empathy for others, and a low tolerance for the ambiguities of human behavior and communication.[14]

This electronic superstitiousness is disconcerting. In my own experience in libraries where I have been involved in training librarians in computer use and in troubleshooting computer problems, baffled librarians and administrators alike have summoned me to clear their screens or bring a downed system back up, addressing me with a mixture of respect and dread that makes me feel like a tribal shaman. Castaneda may relish it, but it alienates me from my professional peers. And it tells me that automation is not working the way it should for us as librarians. We are not comfortable enough around computers, which more and more are becoming an integral part of our profession, perhaps more so than in most other occupations. After all, computer applications in libraries call for much more than word processing. Instead, libraries use computers for the storage, retrieval, and generation of information.

This chapter is not meant to be an exposé of computer phobia among librarians. But the condition does exist, and perhaps to a significant degree. It certainly becomes significant in respect of the extensive use to which computers are being put in libraries.

Not all librarians of course are afflicted with cyberphobia. But many seem to be, and among them are many who are otherwise entirely competent. These are often people-oriented persons who are drawn to librarianship because they take pride in their interpersonal skills and feel compelled to provide an interactive service to people, not machines. This sort of person often does not interact well with machines of any kind, including film projectors, photocopiers, ROM readers, and others commonly found in modern libraries.

Such an observation smacks of stereotyping, but no disrespect is intended. Given the experience of librarians in extracting information from abstruse reference works and our gift for file construction, we should be naturally suited to managing computer data files. We are oriented to data and the diversities of its storage. Unfortunately, too often now a machine and its operation, which have nothing whatsoever to do with the information we are seeking, get in the way. As Ruth Gay says,

> the disembodied nature of computer information, summoned like so many genii out of a bottle, and the interposition of a keyboard between the need and the fact can only strengthen the sense of the elusive nature of the information. Existing only as an electrical impulse, without form until the researcher presses the right keys, its being takes on an almost metaphysical character. Only the right code word will release the treasure. This emphasis on exactness deprives scholars in the humanities of what they call, too lightly, "serendipity."[15]

In her metaphoric way, Gay identifies the major problem to be solved in the field of library automation — making the mechanical storage device truly transparent (as transparent, at least, as a book) for users who are not mechanically inclined. For it appears that the fear of inputting is equitable to a simple lack of mechanical aptitude.

Many people in many professions lack mechanical aptitude, and the lack is fairly prevalent among librarians, who are bookish and not generally given to tinkering. It is almost a truism, of course, that a fondness for tinkering is a prerequisite for success in mechanical aptitude. Gay, however, reminds us that the educational background of most librarians is in the humanities, those disciplines in which "it is hard to see the computer as other than a difficult presence."[16]

This introduces the second method mentioned above for evaluating the level of computer literacy among librarians — the cross-disciplinary approach. In a recent survey of 192,000 college freshmen, 0.0% planned to go to graduate library schools.[17] It seems that students entering the profession are coming from other fields of interest. A discussion with those students, and with established professionals as well, indicates that those fields are (1) the humanities and (2) education. With respect to the humanities, Gay indicates that the research methods in fields that librarians as students of the humanities presumably have learned to favor are fundamentally different from the logic of computer searching. She says,

> The potential problem lies in the difference between what the system is designed to do, which is to find data and retrieve it, and methods of research in the humanities. Here, unlike the sciences, targets are often broad and data ill defined. It is not surprising, therefore, that for nearly two decades the sciences have found computers a natural solution to their needs for current and exact information. . . . But research in the humanities does not usually have the advantage of precision. A narrow search on a computer is not likely to be very enlightening, while a broad one is often wasteful. As the situation now stands there seems to be a basic problem of compatibility between the machine and the scholar.[18]

Inasmuch as a large portion of the library community is drawn from the field of education, a consideration of the reported computer competency of teachers may be meaningful. Clement found a tendency for teachers to have negative attitudes toward computers. He states,

> Most studies indicate that faculty reactions are mixed. Faculty attitudes range from slight interest to open hostility.

Most are indifferent. Experience suggests that implementation problems will be low with teachers comfortable with discussion-oriented instructional modes. For faculty who see their primary teaching role as that of lecturer and information giver, the problems of introducing computers could be severe.

He then adds a pertinent observation that could just as well apply to librarianship:

Putting aside, for a moment, the question of the relative difficulty of an educational degree as compared to other disciplines, one could assume that people are attracted into the instructional society because of a positive exposure to teachers and the teaching art. It is indeed questionable whether these people could be easily persuaded to immediately change the very nature of the career that has attracted them.[19]

According to Stevens, teachers perceive in themselves a lack of computer literacy, a conclusion based on surveys of teachers about to enter the field, experienced teachers, and teacher educators.[20] Griswold reports that education majors have much less favorable attitudes toward computers and a lower understanding of current computer applications than do business majors.[21] He attributes this to curricular differences between schools of business and education. He states, "By the very nature of the curriculum, education students have little opportunity to use computers or to conceptualize how computers might be used to improve their teaching." For business students, however, a "basic understanding of current computer applications is required of all business administration and accounting majors." This requirement is indeed specified in the business school accreditation standards of the American Assembly of Collegiate Schools of Business.[22]

Thus, even in the absence of data drawn directly from the library profession, there is reason to believe that the basal level of computer literacy among librarians is relatively low. Both observation of the behavior of librarians in automated environments and modal attitudes that prevail in disciplines that are major suppliers of librarians support this contention.

Admittedly, these views contradict much of the current literature of librarianship, which tends to read as though computers were invented expressly for libraries and gives the impression that automation will be the basis of all library service of quality in the future. Indeed it seems that the search for the True Integrated System has become the librarian's special quest.[23] And libraries mount enormous "computer literacy" campaigns for their patrons and their staffs, prompting the *Bowker Annual* to report

> Librarians are rapidly losing their technophobia and assimilating as much information about microcomputer applications as they can get. They fill every conference meeting that deals with microcomputers and sign up in droves for trips to local demonstration sites.[24]

Furthermore, library automation courses are proliferating at graduate library schools and are advertised in glowing terms by their promoters.

This flurry of technophilia, however, is less solid than it appears and much less so than is needed by the profession. Literacy campaigns are usually very brief and, therefore, superficial. They are microcomputer oriented. And they often deliver no more to their participants than a certain amount of keyboard familiarity and exposure to a minimal number of software packages, often games or public domain items. Bowker's "conference meetings," for example, are usually workshops on personal computing and do not produce much that is genuinely applicable to the management of large systems.

Most importantly, perhaps, many of the automation courses now being taught in library schools often fall into two categories — microcomputer use and automated applications. Micros are popular because they are inexpensive. But although micros are becoming enormously powerful, a micro definitely is not a mini or a mainframe. And the systems that most library school graduates will confront in the field are based on these much larger machines. Too frequently, moreover, the automation applications courses, which usually concern these larger systems, dwell on the administration of computerized processes and not on the operation and maintenance of the system itself. As a result, many new librarians, who remain our best hope for raising the basal computer literacy level of the profession, are ill-prepared to face the real world of library automation.

AUTOMATED LIBRARY SYSTEMS

While librarians have settled in the domain of the microcomputer, the library has become the domain of the integrated system. Therein lies the literacy gap, the disequilibrium between the librarian's level and brand of computer expertise and the competency necessary to manage a large system. The breadth of this gap becomes more apparent after reviewing the job descriptions reproduced in Appendix C.

These job descriptions were collected from several large public libraries. They all involve areas of computer applications to standard library functions. They cover a wide range of organizational grade levels and an equally wide range of duties. Likewise, the several automated systems to which these job descriptions apply differ from one another considerably and may function quite differently from the systems available across the market today both in hardware configuration and system architecture.

Nevertheless, they indicate a number of important features about library automation. First, most of the positions described do not require any library experience, much less an M.L.S. Only the Computer Systems Librarian position at Public Library E requires library training and experience. The automation requirement in this instance is a "general knowledge of and appreciation for computers and automation." The duties to be performed do not include any real degree of supervision over the system. Instead, they are more on the order of liaison between the professional librarians in the organization and the "appropriate departmental personnel" in the data processing department, namely, the computer operator at a lower classification level, the programmer analyst at the same level, and the director of data processing at a higher level.

It is these last positions that supply the library with the computer expertise not readily available in the profession. And this serves to illustrate the degree to which the profession is separated from computer technology and the degree to which librarians may lack adequate understanding and control of an area of library operations that is becoming increasingly important. As will be discussed in following chapters, this liaison function between librarianship and computer technology becomes crucial to the degree that librarians are unable to master that technology themselves.

Second, the job descriptions reveal that there are in fact various levels and types of computer literacy in a library. For example, the literacy required by the job descriptions is quite different from that required, say, for searching the online catalog. One type of literacy is needed to search online databases, such as DIALOG or BRS, and another to manage the computer fairs and literacy projects that are becoming a standard feature of library public services. Each of these functions requires a different level or type of literacy, and it would be a mistake to assume that expertise in one function automatically produces expertise in others.

Third, the job descriptions indicate the upper level of literacy a manager should expect to have available, from whatever source, to support any automated configuration. This requirement escapes many library managers and even library automation experts who, like Reynolds, believe it is possible for a system to be virtually "turnkey":

> In using a turnkey system, there is little need for hardware or software expertise on the part of the library staff. Some must know more about operating a computer than is necessary for an online service, but that can easily be learned by individuals totally unfamiliar with computer architecture and programming.[25]

More will be said of this later, but mention will be made here of three inaccuracies in Reynold's statement. First, as shown by the job description for library mini-computer operator for Public Library C, many duties must be performed that cannot be "easily learned" by novices. Granted, this library's automated system has some local characteristics, such as card printing, that other systems, especially turnkey systems with online catalogs, may not require. Nevertheless, any computer will require a certain amount of input at the console, and printing devices of one sort or another are a necessity. The duties associated with these and other standard features of the computer may seem elementary to an expert like Reynolds, but in fact they can create difficulties for a novice, even one trained in standard operating procedures by a vendor. The vendor/trainer leaves very soon after installation. During the crucial run phase thereafter, many problems and bugs can develop that a novice will be unable to handle effectively. Aside from being costly in terms of time, money, bother and aggravation, these problems often require a special strategy to undo. A mistake by a non-expert computer operator can cause "muddling," the computerese term for compounding an error by using improper methods to correct it. A computer is a complex machine. When problems occur, diagnosis must be accurate and corrective measures immediate.

In short, Reynold's modesty as an expert misrepresents the complexity of computer handling. A person not adequately trained in computer operations and structure can only perform by rote. If his or her memory lapses, if the system malfunctions, or if a situation occurs that the vendor has neglected to explain while on site, even minor problems become major slowdowns, and they can be complicated greatly by improper operator responses.

The second point, related to the first, concerns the commonly mistaken idea that knowledge of computer programming and architecture is unnecessary to handle turnkey systems. Strictly speaking, this may be true. But as implied in the discussion above, experience in these matters can facilitate management of the system. An experienced operator — even a non-expert — can more easily report problems to the vendor than can others and understand the vendor's instructions, comprehend the written system documentation, train co-workers and simply communicate well about the condition of the system. This readiness can amount to a significant saving in time and money and can minimize slowdowns.

The third of Reynold's inaccuracies is the implication that, once installed, the computer itself is limited to the operations resident in the turnkey software. This is not the case. A computer may well be "dedicated" to a turnkey system. But this does not mean that the computer is "closed" to other programming and even development of modules or functions not available from the vendor. Reynolds states:

> Turnkey systems, like remote access online systems, are generalized. There is sometimes more flexibility in the degree to which a vendor will consider customizing its system to meet an individual library's needs than is the case with a bibliographic utility, but not all vendors are willing to customize. Even if a vendor modifies its generalized software, the library is charged accordingly.[26]

Reynolds is saying that an "off the rack" system will seldom fit all the local needs of a particular library. In this he is correct. For the sake of economy and wide marketability, turnkey systems sacrifice costly flexibility to provide standard functions that will attract the widest range of customers. As a result, a number of locally desired features will be unavailable because the vendor's potential market for them is small. A library possessing the necessary expertise, however, can supplement the vendor's software with locally constructed files and programming, thus creating a hybrid system that will meet the library's total automation needs. This does not mean that the vendor will allow the library to access and modify the turnkey software. But it does mean that a library can create its own modules and then, if so desired and with the permission and cooperation of the vendor (for a fee, of course), integrate or at least link the local products into the turnkey architecture.

Of course, this approach requires that the hardware purchased be of a somewhat larger capacity than might otherwise be necessary. Also, it would require a high level of computer expertise at the library's disposal. Yet it is a highly attractive option that more libraries might advantageously consider. It greatly enhances the flexibility of library automation without incurring the prohibitive costs of developing a total system in-house. Instead, this strategy lets the vendor do what vendors do best, but allows the library to pursue its local specialties as it wishes. In addition, it provides the library with some selectivity in choosing a vendor by permitting the library manager to contract with a vendor that markets the best or most economical versions of the essential library functions, and then to produce the rest locally. The turnkey interfacing can even be made part of the contract. As reported in the case study in Chapter 5, the Smith Library found this strategy suitable to its needs.

In summary, then, the computer literacy requirements of libraries are part of a tiered structure beginning with microcomputer use and database searching and ending with the management and development of much larger systems based on mini and mainframe computers. The structure is not hierarchical, however, because competency at one level does not necessarily indicate competency at other levels. Evidence suggests, furthermore, that the basal computer literacy level for the library profession as a whole may be quite low among the tiers, although some libraries have competency throughout the several levels. As a rule, however, libraries must look outside their own organization and even outside the profession to locate the competency needed to manage large integrated systems. This is true even in the case of so-called turnkey systems, especially if such systems are to be developed to their maximum usefulness and efficiency. The remainder of this study will discuss the impact of these factors on the configuration of automated library systems.

REFERENCES

1. Richard H. Hall, *Organizations: Structure and Process* (Englewood Cliffs, NJ: Prentice-Hall, 1972), p. 122.

2. Amitai Etzioni, *Modern Organizations* (Englewood Cliffs, NJ: Prentice-Hall, 1964), p. 79.

3. Etzioni, p. 87.

4. Marion J. Ball and Sylvia Charp, *Be a Computer Literate* (Morristown, NJ: Creative Computing Press, 1977).

5. Arthur Luehrmann, "Computer Literacy: A National Crisis and a Solution For It," *Byte* 5(7) (July 1980):98-102.

6. D.G. Rawitsch, "The Concept of Computer Literacy," *MAEDS Journal of Educational Computing* 1978(2):1-19.

7. Department of Defense Dependent Schools, *Educational Computing: Support Findings and Student Objectives,* DS Manual 2350.1 (Alexandria, VA: DoDDS, 1982).

8. Tina T. Cheng, Barbara Plake, and Dorothy Jo Stevens, "A Validation Study of the Computer Literacy Examination: Cognitive Aspect," *AEDS Journal* 18(3) (Spring 1985):139-152.

9. Cheng, Plake, and Stevens, p. 140.

10. Neal K. Kaske, "Studies of Online Catalogs, in *Online Catalogs, Online Reference: Converging Trends.* ed. by Brian Aveney and Brett Butler (Chicago, IL: ALA, 1984), pp. 24-27.

11. Richard De Gennaro, "Libraries & Networks in Transition: Problems and Prospects for the 1980's," *Library Journal* 106(10) (May 15, 1981):1045-1049.

12. Craig Brod, *Technostress: The Human Cost of the Computer Revolution* (Reading, MA: Addison-Wesley, 1984), p. 51.

13. "Obscenities in OCLC Data Cost New York State $11,000," *American Libraries* 12(4) (April 1981):176-177.

14. Brod. p. 17.

15. Ruth Gay, "The Machine in the Library," *American Scholar* 49 (Winter 79/80):76.

16. Gay, p. 77.

17. "Freshman Characteristics and Attitudes," *Chronicle of Higher Education* 31(18) (January 15, 1986):35-36.

18. Gay, p. 77.

19. Frank J. Clement, "Affective Considerations in Computer-Based Education," *Educational Technology* 21(4) (April 1981):29.

20. D. J. Stevens, "Educators' Perceptions of Computers in Education: 1979 and 1981," *AEDS Journal* 16(1):1-15.

21. Philip A. Griswold, "Differences Between Education and Business Majors in Their Attitudes About Computers," *AEDS Journal* 18(3) (Spring 1985):131-138.

22. Griswold, pp. 131-132.

23. Susan Baerg Epstein, "Integrated Systems: Dream vs. Reality," *Library Journal* 109 (12) (July 1984):1302-1303.

24. *Bowker Annual of Library & Book Trade Information*, 29th ed. (New York: Bowker, 1985), p. 218.

25. Dennis Reynolds, *Library Automation: Issues and Applications* (New York: Bowker, 1985), p. 218.

26. Reynolds, p. 218.

Chapter 2

Computers and Their Risks

A comparison of the job descriptions (see Appendix) and the DoDDS instructional objectives indicates that optimum use of the kinds of automated systems common in libraries today requires both computer literacy and computer science competencies, as those are defined in the DoDDS document. The case studies from the Phoenix Public Library in Chapter 4 and the Joseph F. Smith Library in Chapter 5 will demonstrate different management strategies for utilizing personnel with the requisite skills for computer operation and maintenance, for as Chapter 1 indicates, there is reason to believe that librarians, while rising in their modal level of computer competencies, may still lack some of the necessary expertise to operate large systems effectively.

This chapter will discuss in greater detail the training needed by library staff members, professional or otherwise, to operate a system based on a mini or mainframe computer. These training needs will be presented in terms of (1) the types of duties and tasks that must be performed to maintain the hardware and software of the system, and (2) the interrelationships that exist between the library and its primary automation vendors, who are the principal source of this training. It is hoped that this method will help library managers and planners unfamiliar with the complexities of system maintenance to visualize in behavioral terms the effects of their decisions on the location and operation of the host computer.

COMPUTER TYPES

Computers are generally classified as micro, mini, and mainframe, depending upon such variable features as processing time, data storage capacity, size and cost.[1] For instance, a mainframe may be capable of processing several million operations per second and of storing hundreds of millions of data characters. A microcomputer, often called a personal computer, home computer, or desktop, processes more in the range of thousands of operations per second and stores significantly less data. Mainframe costs range between a few hundred thousand to several million dollars. Microcomputers may cost up to $10,000.

Minicomputers fall between these two extremes in each of the variables, but in reality the distinct differences between one type and the next are ill-defined, and technological developments are blurring the distinctions even more. For example, some mainframes are constructed in desktop sizes, and microcomputers now in use can process at speeds that once were considered "fast" for minicomputers.

Some crucial differences do exist, however. First, micros are designed to be used by one operator at a time, while the mainframes and minis can accommodate multiple users, sometimes hundreds, simultaneously. Technology is currently being developed to link several micros into a so-called local area network (LAN) that will permit multiple user operations, but large-scale operations of this type are prohibited by the relatively small storage capacity of the micro and by the lack of an effective method to back-up the system, that is, to protect the database from catastrophic loss. In a LAN, one micro becomes a "server" to several others that can access the server's memory and peripherals from remote locations. So while a LAN increases access, it does not really do away with a microcomputer's other limitations. Also, a response time in a LAN arrangement can be unacceptably slow.

Second, mainframes are designed to handle computations and analyze problems that do not generally occur repeatedly in the form of equivalent transactions. That is, programming is tailored to specific problems as they occur. The original concept of the minicomputer, in contrast, was for a small machine that could be pre-programmed, perhaps with "packaged software," and dedicated to some special purpose such as bookkeeping, word processing, or library circulation systems. Indeed, this was the type of machine that seemed most adaptable to the needs of the library even in those early days. Since then, the microcomputer has been developed, and has proven itself well suited to take over much of the packaged type of processing from the mini. Of course, micros do permit local programming, but most personal computer users are more likely to use the software packages that are abundantly available.

Meanwhile, the same technology that gave rise to the microcomputer has dramatically enhanced the sophistication, power, and flexibility of the mini. Not only is the minicomputer suited to host software written for special purposes, but it can also be used to perform other, less typical kinds of functions simultaneously. In essence, the minicomputer provides the capability of dedicating software without also dedicating the hardware. Thus the minicomputer incorporates the best features of both the micro and mainframe options.

At present, many libraries with mini-based systems do not use their machine to fullest advantage. The minis host the vendor-produced software and no more. If the machine has a large storage capacity, half of its effective potential may be wasted by reducing its usage to those narrow objectives for which the minicomputer was designed in the 1960s. This inefficiency can be attributed largely to a lack of computer awareness among librarians and library administrators, who fail to use the machines to their actual capacity.

Third, the three computer types are associated with sets of programming languages that vary in their power to command the machines. Here "power" refers to the configurational characteristics of expressions and the ratio of computer operations executed per expression. Actually, the power of a programming language is largely a function of the governing parameters of the machine it commands. None of the languages bears much resemblance to the language of human discourse. By the same token, the electro-mechanical processing components of the computer do not respond to the programming language directly. They respond instead to mathematical codes which are derived from the programming language by a built-in translator, or "compiler." The situation is much like that of an Englishman who knows no Dutch speaking French to a Dutchman who knows no English.

Translations occur at several removes in a sequence going in both directions.

Given the comparative structural characteristics of micros, minis, and mainframes, languages for microcomputers are not especially powerful at present. Yet they are relatively easy to master and are good for beginners, although many experienced programmers do not recommend learning BASIC. This language, commonly used in one version or another of many microcomputers and in minis also, follows a logic that differs so much from other, more powerful languages that learning BASIC first may hamper one's ability to grasp the other languages later.

The combination of these differences between computer types has profound importance for library automation. According to De Gennaro,

> We need to question the notion that is currently so prevalent in the library field that minicomputer-based systems are somehow more manageable and less expensive than mainframe-based systems. That assertion may have been true in the past when we were implementing turnkey circulation systems, but it probably is not true today when we are dealing with integrated online systems. . . . Many of the reasons for the preference for minicomputer systems in libraries are no longer valid now, as we move toward complex integrated systems which will require the data storage and processing capabilities of powerful and highly reliable mainframes.[2]

Veaner argues rather convincingly against the increased sophistication and inclusiveness of automated library systems to which De Gennaro is referring (see Introduction). But if librarians persist in their demands for bigger and better systems (and it seems inevitable that they will), librarians will find themselves, as De Gennaro aptly observes, pressing ever closer to the blurry upper limits of the minicomputer range into that of the mainframe. Management of these machines requires greater expertise and a competency much different from that now being obtained by so many librarians on the microcomputer level. It may well be that the swelling interest of librarians in micros, as reported in the *Bowker Annual*, is misplaced.

TRAINING NEEDS

Discussion to this point has been guided by a major premise, namely, that a computer should be used to the fullest extent possible. Thus a familiarity with computer languages, system architecture, machine types, etc., are of critical importance. But even if a library decides to use its host computer at the minimum level of efficiency, say, as a residence for a vendor-maintained turnkey system, a certain level of expertise remains necessary for operating mini and mainframe machines. As Susan Baerg Epstein observes,

> it would be wonderful if you could take "turnkey" literally and never have to worry about the system again. In the real world unfortunately, this never happens. A library has to worry about keeping a very delicate piece of equipment operating, making certain that the software is still doing what it originally did, and providing for upgrades and improvements to the operating software. Instead of turning the key on the system and forgetting about it, the library must be concerned with maintenance.[3]

To illustrate the extent of the training needed by a turnkey system manager, this section of the study provides a sample of the duties and responsibilities of the Coordinator of Library Automation at the Smith Library. Although specific duties performed by system managers will vary considerably between libraries and between systems, this selection of duties, it is hoped, will indicate the nature of local maintenance that may be required for optimum performance by a "turnkey" system, whether the system is located in the library or at the central computer facility of the library's parent organization. Although the Smith Library supplements its turnkey system with a considerable number of local files and much programming, the duties to be discussed here pertain specifically to the turnkey hardware and software package supplied by the library's turnkey system vendor.

On page 25 is the specifications overview of the in-house computer at the Smith Library, an Ultimate 2020 minicomputer. The specifications indicate not only the vocabulary and concepts that confront the system manager during operation, but also the types of concerns to be dealt with during the selection process. A librarian, of course, could manage the system without a total familiarity with all the terms, concepts, and specifications mentioned here. Indeed, machine selection is frequently determined by the vendor or the library's governing body as part of the bid process based on the supposed needs of the library. Yet the greater a manager's computer expertise, the more instrumental he or she will be in the system's purchase, management, and configuration, and the more influential in determining exactly what the automated needs of the library really are. And, of course, a thorough understanding of certain of the specifications listed is absolutely necessary, such as disk storage space, port access capability, processing speed, etc.

The Smith system manager is also responsible for the set-up, testing, and operation of a large number and variety of peripherals. These include five types of terminals. Each type is different, and each requires a different procedure for setting such things as baud rate, parity bit type, character code length, character stop bits, etc. These item specific settings are configured by a combination of manual switches and keyboard operations. When problems develop, a certain amount of local diagnostics must be performed in order to determine if the problem is indeed in the terminal or in the central computer hardware or software. If the problem is terminal related, the specific area of trouble must be located — keyboard, display screen, logic board, etc. This information helps the vendor correct the problem more quickly and effectively.

The Smith system also includes four types of printers for letter-quality printing, overdue notices, statistics reports, and screen dumps. The system uses in addition two modems for access to the online catalog from remote locations along with light pens and laser readers for rapid data input. Like the terminals, these peripherals must be set up and matched to the system's parameters. Careful diagnostics testing in case of problems must

FIGURE 1

DESCRIPTION

IMATE's Model 2020 Computer System is a rug-
, high-performance system designed to handle the
st demanding applications easily and efficiently,
do it in a cost-effective manner. The system has
ts base a computer designed by the Digital Equip-
at Corporation, one of the world's top-ranked
iputer manufacturers

Model 2020 is ideal for running concurrent on-
, batch, and time-sharing applications. It can
e as a stand-alone processor with a network of
line terminals for transaction processing, data
management, and time-sharing requirements.

CONSISTENT QUALITY AND DESIGN

he heart of the central processing unit is
IMATE's own operating system — the firmware
supports the application software and governs
internal operation of the hardware. RECALL,
opietary operating language within the operating
em, can be used easily by the novice or experi-
d programmer because it's based on everyday
lish words and phrases. The system also supports
C and PROC computer languages.

Model 2020 is remarkably powerful for a system
s price range. This is achieved by employing
anced multi-processing concepts. By utilizing the
LSI 11 CPU for I/O functions, the ULTIMATE
pheral Processor is free to spend all of its time
uting the ULTIMATE Instruction Set.

TAL PROBLEM-SOLVING FLEXIBILITY

n your ULTIMATE system and appropriate applica-
software, you get information the way you want
nstantly. You are not restricted to regular pro-
mmed reports. Ask virtually any question (using
LL) and so long as the information you want is
where in the system, you get your answer at
, in its most usable form. You have at your fin-
ps all the information you need for creative
lem-solving and decision-making.

the system is completely interactive. You can
rol information in a timely, accurate manner.
ators can enter, revise, and validate information
y. A single entry automatically updates all applic-
records everywhere in the system. You enter
information only once.

FEATURES

ilti-processing capability featuring:
e Digital Equipment Corporation (DEC) LSI 11
ntral processing unit
e ULTIMATE Peripheral Processor especially
signed to execute the ULTIMATE Instructional Set
e 2020 incorporates a unique dual-ported mem-
y design. This improves the memory access time
the ULTIMATE Peripheral Processor. The result is
t ½ times improvement over the Model 2000.
ect addressing of up to 1024K bytes of memory
gh density MOS memory with parity error checking

* Demand paged virtual memory addressing capability of over 300M bytes of storage
* Convenient, attractive, 40-inch cabinet
* 8 to 32 serial ports

SYSTEM OVERVIEW

* Q-bus — The ULTIMATE Computer System 2020
  incorporates a bus architecture design developed by
  DEC. This high-speed, asynchronous, bidirectional
  bus connects the central processor to all of the
  memory and peripheral devices. Each card slot contained in the bus/card cage has four connectors.
  Boards that attach to the bus use either two connectors (dual card = ½ slot) or four connectors
  (quad card = 1 slot). The 2020 system incorporates an eight-slot quad card chassis.
* ULTIMATE Peripheral Processor — The heart of the
  2020 system is a proprietary microprogrammed
  processor board designed specifically for executing
  the ULTIMATE Instruction Set. The highly specialized processor works in parallel with LSI 11 Central
  Processor to support the ULTIMATE operating system in an extremely efficient manner.
* LSI 11 Central Processor — All boot, monitor, and
  I/O functions required for systems operation are
  performed by a standard DEC LSI 11 processor
  board. This processor is also used for running
  maintenance diagnostics.
* Bootstrap — All systems are supplied with a firmware bootstrap that allows either diagnostics/
  hardware debugger or the online system to be
  brought up.
* Memory — 512K-byte MOS memory with parity is
  standard with the ULTIMATE 2020 system. This is
  expandable to 1024 bytes in 512K-byte increments.
  Each 512K bytes occupies one slot.
* Disk Unit — Several Winchester technology 14-inch
  units are available at time of purchase, with unformatted capabilities of 33M bytes, 66M bytes, or
  154M bytes. These high-performance disk drives
  are sealed units that require no preventive
  maintenance. There is a maximum of two disk
  drives per system.
* Tape Unit — A 1600 bpi industry standard, phase-
  encoded tape drive is standard. This auto-loading
  streaming mode device operates at 100 ips or 25
  ips, using ½-inch tape on standard reels.

SYSTEM CONFIGURATION

The system uses slots on the Q-bus for several
options including memory, and both peripheral and
communications controllers. The basic configuration
includes:
* Central processing Unit in an 8 quad-slot Q-bus
  chassis, 40-inch cabinet
* Basic Control Panel
* 512 K-bytes HDMOS of dual ported memory with
  parity error checking
* ULTIMATE Peripheral Processor
* 7 Active Ports
* 1 Serial Printer Port
* 33M byte 14" Disk Drive
* ½ " 1600 bpi Streaming Tape Drive

SPECIFICATIONS

Processor
Word Length:
16 data bits and 19 address bits
Firmware Cycle Time: 165 ns
Firmware Size: 2K locations X 64 bits
Registers: 16 virtual address registers per active
process
Virtual Instructions: 225
Single operand — 90
Double operand — 53
Branch — 57
Shift — 3
Immediate — 10
Input/Output — 7
Generic — 5
Instruction Length:
Variable (1 to 6, 8-bit bytes)
Addressable units:
Bit, byte, word, double word, triple word

Q-bus
Type: Asynchronous, bidirectional
Bus bandwidth: 1M byte/sec
Slots: 8 quad slots
Data Lines: 16
Address Lines: 19
Serial Ports: 8 to 32

Disk
Winchester-Type Disk (14 in.)
33M bytes, 66M bytes, or 154 bytes (unformatted)
Capacity: 30M bytes, 60M bytes, 140 bytes
(formatted)
Track-to-Track Seek: 8 ms
Average Seek Time: 45 ms
Maximum Seek Time: 85 ms
Average Latency: 9.7 ms
Transfer Rate: 1.04M bytes/sec.
MTBF: 8,000 (power on hours)
MTTR: 30 (minutes)

System Specifications
Power Requirements: 115 V/60 Hz @ 10 A or 230 V/
50 @ 5 A
Cooling Requirements: Standard temperature con-
trolled office environment
Space Requirements: Minimum 2 foot air circulation
clearance on each side
Height — 40 in.
Width — 24 in.
Length — 36 in.
Weight: 400 lb. (approx.)
Environmental Characteristics:
65°F-80°F @ 20 % -80 % relative humidity

The information and specifications in this document
are subject to change without notice. This document
contains information about ULTIMATE products or
services that may not be available outside the United
States. Consult your authorized ULTIMATE Dealer.

One of the ULTIMATE family of advanced computer products available through authorized dealers worldwide.

THE ULTIMATE CORP.

Corporate offices: 77 Brant Avenue, Clark, New Jersey 07066 (201) 388-8800 also be performed in a fashion similar to that used in maintaining the terminals.

A certain familiarity with microcomputers and various types and brands of microcomputer software packages is also useful because interfacing of these machines with the Ultimate 2020 is common. For instance, remote users access the online catalog via the modems using their micros as terminal emulators. Also, the Dynix system utilizes a microcomputer dedicated with special software to establish an OCLC interface.

These different brands and types of peripherals exist at the Smith Library because automation has been underway there for several years. Few turnkey systems, of course, will begin with such a large array of equipment types and brands. Most system managers will probably initiate their automation projects fully intending to standardize peripheral equipment completely. After a period of time, however, some diversification is inevitable. As terminals and printers are replaced, or as the system grows, newer, less expensive and more sophisticated pieces of equipment are sure to be purchased, thus requiring the manager to have an understanding of peripheral operation so that the newer items can be linked into the system.

Supplies acquisition for the system are not difficult, but requires an awareness of many types and qualities of products. These include different ribbons for each type of printer, printer paper for various types of printouts, and tapes for the tape drive. Also needed are assorted screwdrivers, pliers, and a soldering gun for switching cables and making any minor wiring repairs that are necessary.

The major mechanical function related to the computer itself is operating the tape drive. This is used in five different processes. The first is the back-up, or file-save process, which periodically stores on tape the entire contents of the computer disk. This is a lengthy process that can be disruptive to and disrupted by normal system operations, if performed simultaneously. The back-up process, therefore, must be performed after normal working hours, when no one is using the system. All access lines must be blocked to prevent dial-in access, and during the file-save, the screen is monitored to check for group format errors (GFE), which in essence are errors in the data linkage system. The back-up is necessary because it will allow the system and the files to be restored if the data on the disk is inadvertently lost or if the disk is damaged as a result of fire, head crash, sabotage or whatever.

The tape drive at the Smith Library also runs the Transaction Assurance Program (TAP), which is a standard feature of the Dynix system. This program (1) processes the daily transactions performed on the system and transfers them to the disk, and (2) provides an interim record on tape of all transactions to supplement the back-up function between file-saves. Thus, in case of damage to the computer, the TAP tape, along with the most recent file-save tapes, can accurately restore the files onto a new disk up to the precise moment the damage occurred.

The third tape function is the file restore, which is the process that must be followed in the unlikely event that damage occurs to the disk, thus necessitating a replacement and a rebuilding of the files on the new disk. More frequently, however, the file restore is used to resize the files, clear group format errors, and reconfigure the system after upgrades to the computer. Both of these processes must be clearly understood by the system manager.

The fourth and fifth functions performed by tape is the loading of new releases from Dynix and from the Ultimate Corporation. The Dynix upgrades are sent to the library on tape and are loaded on the drive at a specified time, usually after hours. Once online, Dynix programmers dial in to the Smith system and complete the loading of the new release before the library reopens. New releases from the Ultimate Corporation, however, must be loaded and installed by library personnel. It is a lengthy process, taking several hours, and one for which documentation is supplied outlining a sequence of steps that must be followed exactly. Although the documentation is detailed, it is seldom lucid and often confusing.

These diverse tape functions require a substantial tape library, some of which must be stored off site to ensure file security in case of a major catastrophe at the library. The tapes must be constantly rotated, inspected for parity errors, and replaced if defective. The tape heads on the drive must be cleaned periodically, along with the printer.

Obviously the system manager must be immune to the fear both of machines and inputting. Aside from following computer logic, an ability to read schematics, test wiring and perform simple mechanical operations are valuable assets. The system manager must be able to shut down the computer (a "warm stop") so that the operations listed above can be performed, then restart it (a warm or "cold" start) and bring the system back online. Programs interrupted to perform such operations must be restarted, and this occasionally requires a complicated diagnostics procedure.

With respect to the online operations, the Smith Library system requires two levels of maintenance, both of which are beyond that of the normal inputting of circulation, cataloging, and other transaction data. The first level involves correcting certain problems such as books that won't check in, fines that won't clear and operator errors. Some of these can be corrected through normal operator input procedures by using them in a slightly different way. The system manager must understand how to use this reverse logic and devise strategies for doing so on a case by case basis. The most frequent occurrences of this type of maintenance at the Smith Library are in circulation, TAP, and online catalog functions.

Frequently, however, these and other problems can only be corrected at the terminal control level (TCL), which directly accesses the files and records that hold the data and the system parameters used by the turnkey software. These are items that the normal user never sees because they are one level closer, so to speak, to the operational control features of the computer itself. At TCL, the system manager has access to all the data in the computer and can study the complex linkages between different portions of that data. By manipulating the appropriate portions and tracing the linkages across several files that may be involved, the system manager can undo nagging problems and erase their residual effects. Of course, this requires a thorough knowledge of the computer's operating system and of the structure of the files stored in the system.

TCL is also the level at which programming is written and files are created. None of this is necessary, strictly speaking, as part of the turnkey functions. The Ultimate's query language, however, known as Recall, is also input at TCL. This "language" is actually a set of commands used to construct near-English sentences that generate on the spot highly accurate and current reports of system conditions and file contents. These reports can be displayed on the terminal screen or sent to the printer. Because the Recall reports can be made very specific and detailed, they are invaluable for problem solving and file maintenance. The effective use of Recall, however, requires close familiarity with the structure of files and records within the database. Needless to say, the system manager must be highly proficient in the query language and the system architecture that allows its use.

These, then, are a few of the maintenance functions required by a so-called turnkey system. As Epstein implies, the term is a misnomer. Like "user friendly," "integrated system," "authority control," and a host of other catchy phrases, it is vendor-defined and is not necessarily accurate. The types of operations discussed above need to be performed on the system whether by trained computer operators at a central facility or by library staff members, professional or otherwise, in the library. Before making the decision to locate the host in house, the library manager should consider carefully where the necessary expertise to run the system will be found.

VENDORS AND TRAINING

Obviously, the manager of an automated system, even a turnkey, must possess a number of machine skills. This is indicated by the collection of job descriptions in the Appendix, and verified by the informal "job audit" above of the system manager at the Smith Library. Included in these skills is a thorough understanding of the system architecture and the structure of files, records and linkages. Furthermore, the manager must be able to recognize errors and trace them almost intuitively, and then communicate them in the language of the vendor.

These kinds of skills come only with online experience. Vendors do not provide them, despite aggressive claims to the contrary. What is needed is outright re-education, and vendors are not in the re-education business. Instead, they provide limited training in three phases, the first being the pre-installation phase. During this phase, the vendor sends advance documentation to the library to prepare the staff and physical plant for the installation. Worksheets may be included on which the library staff specifies the parameters and limitations to be set up in the system. These parameters include such things as patron types, item types, loan periods, fine schedules, collection codes, material type codes, and statistics structures. Although the library staff deals with these parameters on a daily basis, they can be deceptively difficult to enumerate and prioritize on the spot. Hence the advisability of assembling this information during the pre-installation phase. Doing it in advance can save valuable time when the installation team arrives.

The installation phase is the vendor's primary training vehicle, and usually it is highly insufficient. The time allotted is generally too short, depending on the amount of training time a library can afford to purchase. As a result, routines are demonstrated almost mechanically at the expense of conceptual understanding. Fortunately, the training is usually "hands on," with each participant given a good dose of online time. But normally only a handful of library staff members, usually supervisors, receive the training, again depending upon what the library can afford. These participants are then expected to train their own departments in the intricacies of the system after the installation phase is completed, and after the vendors are no longer physically available to clarify procedures and answer the questions the supervisors may not have thought to ask.

During this training phase, a grueling amount of information about totally new methods of seeing to one's habitual duties is force-fed to a relative few during a relatively short period of time. The result in many cases is training without learning. Participants often know what steps to perform, but they don't know why. They act as if in a ritual performance, which is an unsound way to approach a machine. Panko observes that a lack of good training will result in "crippled use and nonadoption" of the system's operating principles. He gives the following as minimums for effective vendor training during the installation phase:

1. Two days per application package.
2. A half-day to a day for computer and operating instructions.
3. Casual users need refreshing.

4. Advanced users need advanced modules.
5. Background hardware knowledge for intuitive use.
6. Background software knowledge for intuitive use.
7. Concepts of data files and data bases.
8. Knowledge application development needs.
9. Knowledge of proper management (back-up, passwords, etc.).[4]

Even when a vendor addresses these items during the installation phase, the coverage is generally rushed and sparsely distributed, especially at the levels of training needed by the system manager.

The post-installation phase consists of vendor-hosted workshops, correspondence and telephone conversations. The workshops, like the sessions during the installation phase, are usually intense. They are also infrequent, perhaps yearly. They are costly, crowded, far away, and brief. As a result, they, too, are short on effectiveness.

The correspondence may consist of vendor newsletters that are more hype than instruction; system "tips" that may not be of any use at all since they are blanket mailings to all customers; and revisions of documentation or documentation about new releases. A word about documentation is in order. It is notoriously arduous, especially that accompanying hardware. Written by machine people, it is highly technical, conceptually elliptical and often ungrammatical. The documentation supplied by turnkey vendors for their own software tends to be much better, but nearly all automation documentation is extremely lengthy and often deadly in its technicalities. Yet it must be mastered by the system manager virtually at once.

As the case studies will show, telephone conversations often are the only way to keep the system up during crisis situations in the post-installation phase, and the system manager should be prepared for long hours on the line. All problems should be carefully documented as they occur in order to make these telephone conversations as productive as possible. The vendor will need to know the inputs, responses and results of every problem brought to his or her attention. No solution is possible without full details. It is a tedious matter, even for someone who has a background knowledge of computers, data files, and system architecture. For someone who does not, the prospects are even less favorable.

According to Panko, this phase of client/vendor relations is extremely important. Vendors, he says, should be prepared to provide "ongoing hand holding" because

1. People do not learn everything in class.
2. People forget.
3. Some people need advanced capabilities.
4. People compound errors into "muddles."[5]

Not all vendors, however, follow Panko's advice, and the library should not expect much hand holding.

In an automated environment, the library is very much in the middle of a difficult situation, surrounded by the software vendor, the hardware manufacturers, and the hardware maintenance contractors. These are commonly three different vendors, or sets of vendors, with whom the library must deal. It would be nice if between them they would provide the library with the re-education necessary to ensure the optimum operation of the system upon installation. They do not. As a result, the library must provide the compensating expertise from whatever source is at hand. The whereabouts of the expertise is the real determining factor underlying the issue of the whereabouts of the host computer, but few library administrators realize this prior to making the decision to locate their computer in house.

REFERENCES

1. See H.L. Capron and Brian K. Williams, *Computers and Data Processing*, 2nd ed. (Menlo Park, CA: Banjamin/Cummings, 1984), pp. 270-275, for a fuller discussion of computer types. For an excellent discussion of computer types and the limitations of the micro in libraries, see David H. Carlson, "The Perils of Personals: Microcomputers in Libraries," *Library Journal* 110(2) (February 1, 1985):50-55.

2. Richard De Gennaro, "Integrated Online Library Systems: Perspectives, Perceptions, & Practicalities," *Library Journal* 110(2) (February 1, 1985):39.

3. Susan Baerg Epstein, "Maintenance of Automated Library Systems," *Library Journal* 108(22) (December 15, 1983):2312-2313.

4. Raymond Panko, "End User Computing and Information Centers," from Info-Tech Seminars, Management Programs of Hawaii, Honolulu, HI, January 29-30, 1985.

5. Panko.

Chapter 3

Remote Host or In-House?

As discussed in the previous chapters, the premises of this book may be summarized as follows. First, computers that host library systems should be used at their maximum level of effectiveness, which generally is significantly greater than found in those functions supplied by a turnkey vendor. These maximum levels should be determined by (1) the needs and specialties of the library, (2) the availability of expertise to develop the system, and (3) the characteristics of the computer itself. To achieve the desired level of development, it must be added, requires a high degree of resident competency in computer literacy and computer science.

Second, librarians as a group lack the necessary expertise at lower levels of efficiency characteristic of the turnkey software that has wide applicability in libraries, but low specificity for any one library. Although it cannot be demonstrated conclusively, the behavior of librarians in an automated environment and the nature of their educational backgrounds suggest that in general they lack computer expertise, especially with regard to the larger minis and mainframes of the type required for sophisticated systems.

Third, to support such powerful systems, and especially to improve them, libraries may well be forced to go outside of the profession to find the necessary expertise to operate the systems. This could have two negative effects on the profession. First, it could fragment the profession because librarians would lose control over a major portion of their own work. Second, simply because this expertise will be supplied by nonlibrarians, the applicability of the systems they design and operate could be compromised by their lack of training in librarianship.

From these premises it is appropriate now to draw some preliminary conclusions about library automation as it affects the decision as to where to locate the host computer. The hope is that these preliminary findings will be helpful in decision making. An attempt is made in the following chapters to refine them to a more acceptable degree. Here, however, they serve as focal points for evaluating the following case studies, which embody many, if not all, of the considerations discussed up to this point.

STAFFING

The staffing of automation projects generally follows one of two patterns. First, a separate department is organized and given complete and exclusive responsibility for the system. This allows the automation staff to concentrate on maintaining the system and provides the organizational and budgetary resources for the division's operation. As indicated by the job descriptions in the Appendix, it is common for large libraries to create such departments and separate them from other parts of the library. This is especially true if the automation staff is composed of nonlibrarians. But this approach creates yet another entity in an organization that is already highly departmentalized, and is almost certain to impede communication within the organization. In an earlier study I discussed the importance of proper management behaviors in the context of library reorganization.[1] It is enough here to say that a library's goal structures, organizational form and staff involvement are critically important in any type of reorganization, including the planning and developing of an automation project.

The case study from the Smith Library at Brigham Young University-Hawaii Campus illustrates an instance in which an automation project resulted in the formation of a separate automation department within the larger library organization. After several years of developing a database on a computer in the university's computer center and designing its own automated system on that computer, the library chose to purchase its own computer, install on it a limited turnkey integrated system, and create a separate automation department within the library staffed by personnel with sufficient expertise to manage the turnkey system and develop other features as these were needed. Although this arrangement has worked well for the Smith Library, it has not developed without a good deal of difficulty and occasional frustration, both with regard to operation of the system itself and to the establishment of cooperation between the new department and other functional areas of the library and the university.

The second method is to assign the responsibility for the automated system to a department in the library's existing organizational structure. Additional staff members, if any, are then attached to that department under the direction of the current supervisors. This method may be more healthy for organizational communication, but it risks insufficient organizational and budgetary support for the automation project, inasmuch as resources for the new functions may be expected to come out of the existing allocations in the department affected.

The departments most often designated to supervise the automation project under this approach are the technical services or the circulation departments. The first is the likely choice if the system includes integrated modules such as cataloging and acquisitions that are based on centralized bibliographic authority files. The circulation department may be a good choice if the system is basically intended for circulation control.

This staffing pattern is common in system configurations in which the host is located at a remote computer center. In such a case, the addition of computer literate personnel to the staff may well be judged unnecessary. Instead, current staff members are merely assigned additional responsibilities that include the appropriate system-related tasks. The discussions in Chapters 1 and 2 suggest the potential dangers of such a decision.

The Phoenix Public Library case study demonstrates an instance of this approach. The technical services department was given responsibility for the ULISYS system. This was a circulation system with a few searching capabilities, and so could serve as a limited online catalog. The computer used by Phoenix Public was located at the city's central computer facility, where it was operated by personnel with no library experience, and no new personnel were added to the library staff. Automation training for the existing staff was minimal, and system monitoring duties were simply added to the duties of the supervisors in the technical services department. This case study shows clearly the weaknesses of such an arrangement. And yet the study also illustrates how this approach can become successful for libraries lacking computer experience.

In either of the two approaches, it is obvious that someone must be in charge of the library's automation project, and that this person must be an official member of the library organization. Ideally, the person will be a professional librarian who is competent in computer technology and operation. Such a person would not only be familiar with the functions and operations of a library, but would also be able to apply the logic and mechanics of sophisticated mini or mainframe computer procedures to those functions. Because such doubly qualified personnel are relatively few in comparison with the number of libraries embarking on large-scale automation projects, a library will more likely face the decision of upgrading the computer expertise of a professional librarian or teaching a programmer/analyst the necessary aspects of librarianship that will enable him or her to operate the system in accordance with established library procedures and philosophy.

Either approach involves a great deal of supplemental training for the person involved, and each library presents peculiar conditions upon which the decision must be based. Peter Drucker gives some advice that may be meaningful in these circumstances. He says:

> You need someone who has been through a few troubles and doesn't panic . . . and knows that the basic rule of new systems is that everything degenerates into work. Where are you going to get such a person? Not from the outside, because you can't afford to take that risk. You hire people from the outside for work you understand so that you can help them when they get into trouble. No, you seek such a person inside the organization, and once identified you have to free him or her to take on new responsibilities. That

means you have to be willing to slough off or downgrade or deemphasize something else.[2]

Although Drucker here was not referring specifically to automation, he was addressing a conference of librarians. And although he was speaking on the organizational and not the professional level, his advice is nonetheless applicable in the present situation. He is saying that librarians should rely on expertise from outside the profession only in matters that are reasonably understood already, such as, say, bookkeeping, legal, special projects, etc. In matters where preliminary expertise within the profession is lacking, upgrading from within is the advisable option.

For the individual library wishing to take Drucker fully at his word and upgrade a present staff member to manage an automation project, either in-house or remote host, this almost necessarily includes formal classwork and self-instruction. Commercial workshops are numerous and generally informative, but they are expensive and limited in their coverage. As mentioned in Chapter 2, vendors are oriented toward limited customer training, not toward participant re-education.

For the profession, this upgrading from within has broad, long-term implications that will be discussed more fully in Chapter 7. In essence, it means ensuring that in the near future the profession will have a good supply of librarians with the computer expertise to manage the expanding number of online library systems, and to contribute directly to the development of new software and hardware that will be fully compatible with the philosophy of professional librarianship.

HOST LOCATION

We come now to the focal point of the study: In light of the considerations discussed to this point, should the library choose to locate the host computer in house where it will be maintained by library personnel, be they librarians, programmers, or some combination thereof? or should the computer instead be located at a centralized computer facility belonging to the library's parent organization where the computer will be maintained and operated by computer specialists who, though experts in automation, are not librarians? Library literature is not abundant on this topic, but a few sources do provide some pertinent views. According to Reynolds, one advantage of the in-house configuration over remote-access configurations is that

> the library may be able to exercise a greater amount of control over certain types of decisions regarding hardware performance and upgrade. The system still goes down on occasion and there still are periods when response time is unacceptably slow. But when the library owns the computer, at least it will usually be able to deal more directly with the appropriate maintenance agency, rather than having to simply notify an online service or a computer center, only perhaps to be then left in the dark about the nature of the problem, what is being done about it, or how soon it will be corrected. Being better informed does not necessarily solve the problem more quickly, but it does at least enable the library to have a more direct grasp on the situation.
>
> A library that has a turnkey system also generally has more direct control over decisions about switching to new equipment or upgrading existing configurations than is possible with online services or local computer centers. The degree of control over software enhancements is about on par with an online service, with similar opportunities to those available for users of online services: namely, through participation in channels that forward user groups and other advisory suggestions to the vendor.[3]

The case study from the Phoenix Public Library validates Reynold's views to a point. Nevertheless, dealing more directly with the appropriate maintenance agency, as he puts it, does not mean that the problem will be corrected any more quickly than it would otherwise. On the contrary, if the library staff cannot communicate the nature of the problem clearly enough to help the maintenance agency quickly locate the source of the difficulty, correction of the problem will be delayed. Similarly, lack of computer expertise may prevent the library staff from understanding the nature of the problem, whether the computer is in house or not.

Like Reynolds, Corbin claims that in-house location will give the library "complete control" of its system. The only disadvantage, he says, is that the library will be required to provide a site that meets the temperature, humidity, and power requirements of the computer.[4] Conversely, he finds that the only advantage of a remote host configuration is that the library will be spared the need to provide the site. The disadvantages are (1) "the library might feel a loss of autonomy when it must depend upon another unit for its computing power," and (2) "few computing center staff . . . understand the complexities of library operations and files. This lack of understanding can result in serious problems for the library and major deficiencies in the effectiveness of the service operations."[5]

Elsewhere, Corbin addresses at some length the complexity of knowledge and the depth of training and experience needed by the library staff members charged with operating the computer and managing the system.[6] Yet here he curiously omits the need for providing such expertise among the staff as a disadvantage, while summarily minimizing the ability of programmers and systems analysts to understand library complexities. If the latter condition were really true, there would be no library automation systems even now. Library automation has reached its current level of development because librarians and computer experts are in fact able to work well together under the appropriate circumstances.

So although Reynolds and Corbin argue that the in-house location of the computer gives the library better control over its own resources, in reality the advantages may be compromised by lack of computer skills in the library. On the other hand, the disadvantages they attribute to the remote host configuration can be overcome by effective interagency communication, as is discussed below by Genaway. There is a more fundamental issue involved here, namely, the ability of the library to administer its automation resources effectively regardless of where the host computer is located. Reynolds and Corbin ignore the fact that such ability is not always available.

The *Library Systems Newsletter* reports that

> one area which frequently causes concern is the decision as
> to whether the library should "go it alone" with a single
> stand-alone system or whether it should seek to combine
> with a group of libraries in the purchase of a shared system.
> Many factors have a bearing on the decision: local politics,
> funding sources, the physical location and commonality of
> interests of the libraries, personality factors, and cost. The
> mix and weight of these elements will vary from one situa-
> tion to another. However, cost is always an important con-
> cern.[7]

This introduces a variation of the remote host option—the
shared system. The present study will not specifically address
this configuration, but the considerations in this arrangement are
similar to those discussed here. And as *LSN* points out, there are
numerous factors to be considered, although no mention of computer literacy requirements is made. Instead, the emphasis is on
cost. The editors go on to outline some "rules of thumb" by
which libraries can calculate the comparative costs of the in-
house and shared system options. Much of the discussion centers
about telecommunications costs which, of course, will be somewhat higher in a remote host than in an in-house configuration.
Yet it must be remembered that the telecommunications costs
will be high for any type of system designed to serve a large
library with many branches. The difference between linking the
branches to the host computer at the main library and those of
establishing links to the central computer facility may not be significant enough to override the consideration of available computer skills, the lack of which can also be costly.

Coyle also addresses the issue of costs in her discussion of host
location. Amortizing turnkey installation costs and the costs of
"timesharing" over a period of five years, she states that "it will
be more expensive to timeshare as you reach and exceed five
years, but the short run may be cheaper and safer, especially if
you do not want to commit your organization to a system indefi-
nitely."[8] Coyle here introduces yet another remote host varia-

tion, that of timesharing, which means that outright purchase of a computer by the library may not be necessary. Instead the library may be able to buy time and space on the vendor's computer or on a computer already installed in the computer center of the parent organization. The contractual arrangements of such a configuration will vary widely between situations, but Coyle discusses several advantages of this approach. First, the library's initial cost will be much lower since a computer purchase will not be necessary. Second, hardware concerns will be minimized, as will maintenance costs, because these will be handled largely by the computer center staff. Third, the implementation of the system will be somewhat expedited because the core elements of the system are already in place. Like Reynolds, she finds the chief disadvantage to be that of limited control over operation and development of the system.

Coyle does not address the issue of computer skills. She seems to assume these skills to be sufficient for system operation and development, and implies that hardware difficulties are more severe in an in-house system than in a timesharing system. But in actuality, system operation and hardware problems are aspects of the same issue — the presence or not of the skills necessary to control them.

Like the editors of *LSN*, Genaway addresses an aspect of consortium timesharing. He states that among the participants, "policies need to be developed that are mutually acceptable to all members of a multi-jurisdictional consortium. Also, all members need to be aware of each other's procedures and policies."[9]

Although these remarks are directed toward consortia, they represent precisely the type of strategy that, if applied between a library and a computer center, can help overcome to a large degree the disadvantages of locating the host computer at the center instead of on the library premises. If the disadvantages of not locating in house can be overcome through effective interagency cooperation and communication, perhaps through cross-training of personnel from the library and the computer center, the lack of computer expertise among librarians may be considered less damaging than it might otherwise be. And if the computer must be run by non-librarians to be effectively utilized, the advantage of locating in house all but disappears. Admittedly, immediate physical access to the computer is assured by locating in house, and telecommunications costs may be reduced, but if the computer is staffed by non-librarians, immediate access is of little advantage. The salaries of the computer specialists would probably cost the library at least as much as buying the necessary services from the computer center, and the specialists would be difficult to supervise because the supervisor might be unable to evaluate their performance fairly.

The disadvantages of the remote host, to reiterate Reynolds, Corbin, and Coyle, are loss of system control in case of problems and loss of decision-making power with respect to system operation and development. The Denver Public Library, however, recently found the opposite to be the case. After several years with an in-house system supplied and maintained by a vendor, Denver Public and the other members of the Colorado Alliance of Research Libraries (CARL) made the decision to proceed with local development of a shared integrated system. With respect to Denver Public, this decision was based on a desire to obtain greater control of their system even though it involved working through the consortium, and even though the host computer was to be located at a remote site. In this case, the issue of host location properly became secondary to the larger issues of system performance and control which administrators at Denver Public felt had been lacking in the previous in-house system.

The ongoing success of the CARL venture suggests that proper organizational communication and cooperation, just as Genaway observed, can minimize the disadvantages that Reynolds, Coyle, and Corbin find in the remote host configuration. Of course, the simplicity of this statement belies the difficulty that would be involved. Intergovernmental cooperation is not easily achieved. As I have said elsewhere,

> If cooperation between libraries is difficult, cooperation between government agencies can be even more trying. The lack of uniformity in procedures, materials, resources, and clienteles across a number of agencies can cause coopera-

tive efforts to backfire. When problems arise, as they inevitably will, the home agency will blame those on the other end.[10]

And when this occurs, as Parker says, "a number of communication channels begin to shut down without fanfare."[11] But librarians may be more adept at communication than automation, at least for the present. The case studies that follow are intended to help library managers make the important decision concerning the in-house/remote host choice.

REFERENCES

1. T. D. Webb, *Reorganization in the Public Library* (Phoenix, AZ: Oryx Press, 1985).

2. Peter F. Drucker, "Managing the Public Service Institution," *College & Research Libraries* 37 (January 1976):8.

3. Dennis Reynolds, *Library Automation: Issues and Applications* (New York: Bowker, 1985), p. 219.

4. John Corbin, *Managing the Library Automation Project* (Phoenix, AZ: Oryx Press, 1985), p. 21.

5. Corbin, p. 21.

6. Corbin, pp. 159-168.

7. "Stand-alone or Shared? Costings and Considerations in Turnkey Configurations," *Library Systems Newsletter* 3(4) (April 1983):25-26.

8. Mary L. Coyle, "The Integrated Library in a Timesharing Environment," in *Conference On Integrated Online Library Systems, September 26-27, 1983: Proceedings,* ed. by David C. Genaway, rev. ed. (Canfield, OH: Genaway & Associates, 1984), p. 37.

9. David C. Genaway, *Integrated Online Library Systems: Principles, Planning, and Implementation* (White Plains, NY: Knowledge Industry Publications, 1984), p. 19.

10. Webb, p. 47.

11. Thomas F. Parker, "Resource Sharing From the Inside Out: Reflections On the Organizational Nature of Library Networks," *Library Resources & Technical Services* 19 (Fall 1975):353.

Chapter 4

The Phoenix Public Library: Introduction to the Case Study

Each case study presented here focuses on those operational matters that are most pertinent to the automated configuration of the library involved. Because the Phoenix Public Library utilizes the remote host configuration, the study in this chapter will focus on the communicational and cooperational requirements Phoenix Public faced with respect to the ULISYS vendors, the hardware maintenance contractors, and especially the personnel at the city's Management Information Systems (MIS), where the computer was located.

In the remote host arrangement, interagency communication is of the utmost importance. It requires political, governmental, and organizational acumen on the part of the library administration, as well as personalities suited to cooperation that is at once amicable yet determined. The library in this situation must access its system and communicate with the vendors, who can be quite remote themselves in every sense of the word, through the rather inflexible intermediary of operators and programmers at the host location who almost certainly will have no initial library training.

Yet a library without the necessary technical expertise to operate its own computer may find this a favorable arrangement, especially if the library administrators possess the political and interpersonal finesse necessary to work well with other agencies in the parent organization. The remote host configuration, in fact, has some unsuspected political advantages insofar as the public library is concerned. In the first place, a library configured in this manner will be able to marshall more economic and political leverage from the parent organization in periods of vendor/client crisis than will a library "going it alone" with an in-house system, a fact that will not be lost on the hardware and software vendors. Secondly, the remote host arrangement brings the library into the larger arena of the local governing structures where it belongs and where it needs to be, if it is to survive.

The library has long suffered from what Getz calls a "semi-autonomous character"[1] which derives from two conditions. First, at least in the public library, the users are largely middle- and upper-class individuals with higher-than-average levels of income and education. This somewhat narrow political base, according to Getz, lessens the library's bargaining power in the government bureaucracy, which fact he validates by demonstrating the readiness of municipal budget makers to alter library budgets radically and arbitrarily with little or no regard to the actual effectiveness of the library itself.[2]

Second, while the administrators of other agencies of the parent organization have degrees in public or business administration or a variety of other fields, library managers are conspicuously library scientists with common professional training and backgrounds. This provides great unity among library administrators, but it alienates them from their counterparts in other agencies and relegates the library to the fringes of public administration. As Getz observes

> While these organizational characteristics may enhance the quality of library services and lead to more rapid diffusion of innovations than is common in other local government services, the cost may be a lack of political success.[3]

Similarly, Lowell Martin warns that libraries must avoid their customary "institutional provincialism and think more in terms of library users and functions."[4] He says also, "Our taxonomy has an institutional bias. . . . We have not only the separateness of libraries to bridge but also the separateness of these discrete parts of our governmental-educational-economic fabric."[5] If properly handled, the interrelationships that arise from a remote host configuration can help integrate the library more firmly into the workings of the parent organization. Automation, which is becoming an increasingly common feature at all levels of interorganizational structures, can serve as a bridge to alleviate some of the library's chronic separateness.

The Phoenix Public Library presents a particularly meaningful demonstration of the importance of interagency cooperation, and not only with respect to the remote host configuration. Shortly after the installation of the ULISYS automated circulation system, the library, which previously had been a separate department of city government, was consolidated with the city's Parks and Recreation Department. The new entity was labeled the Parks, Recreation, and Library Department (PRL). The library director immediately became a PRL assistant director and lost direct contact with the City Council. From that point on, all library planning had to be approved by a department director who admittedly knew nothing about libraries. Parks and Recreation personnel began to frequent the library giving directions and taking over functions previously handled by library staff members. Tables 4-1 and 4-2 illustrate the present organizational positions of the Phoenix Parks, Recreation and Library Department, the Phoenix Public Library, and Management Information Systems in the municipal government structure.[6]

Thus the Phoenix Public Library is required to cooperate with other government agencies on many levels and almost continuously. As the parent organizations of other libraries are forced to implement budget-cutting strategies, Phoenix Public may well be an example of future trends for many libraries. The prospect of interagency consolidation is not without its fearful side, but it has the advantage of placing the library more in the mainstream of the governing operations of the parent organization.

The lessons to be learned from the failures and successes of the remote host configuration at Phoenix Public, as reported in the following case study, may have implications beyond library automation. For many libraries, as for Phoenix Public, the interagency cooperation made necessary by a remote host configuration perhaps should be seen as a prelude to even more extensive organizational interaction. Drawing on the automation expertise and resources resident elsewhere in the parent organization may be the initial phase of the diminishing of the library's semi-autonomous character and institutional provincialism. A shrewd li-

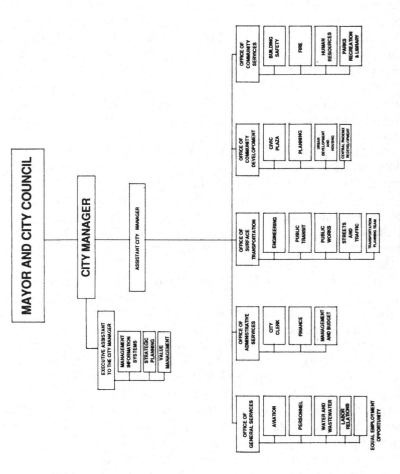

MAYOR AND CITY COUNCIL

CITY MANAGER

ASSISTANT CITY MANAGER

EXECUTIVE ASSISTANT TO THE CITY MANAGER
- MANAGEMENT INFORMATION SYSTEMS
- STRATEGIC PLANNING
- VALUE MANAGEMENT

OFFICE OF GENERAL SERVICES
- AVIATION
- PERSONNEL
- WATER AND WASTEWATER
- LABOR RELATIONS
- EQUAL EMPLOYMENT OPPORTUNITY

OFFICE OF ADMINISTRATIVE SERVICES
- CITY CLERK
- FINANCE
- MANAGEMENT AND BUDGET

OFFICE OF SURFACE TRANSPORTATION
- ENGINEERING
- PUBLIC TRANSIT
- PUBLIC WORKS
- STREETS AND TRAFFIC
- TRANSPORTATION PLANNING TEAM

OFFICE OF COMMUNITY DEVELOPMENT
- CIVIC PLAZA
- PLANNING
- URBAN DEVELOPMENT AND HOUSING
- CENTRAL PHOENIX REDEVELOPMENT

OFFICE OF COMMUNITY SERVICES
- BUILDING SAFETY
- FIRE
- HUMAN RESOURCES
- PARKS RECREATION & LIBRARY

TABLE 4-1

48

PHOENIX DEPARTMENT
OF PARKS, RECREATION AND LIBRARY

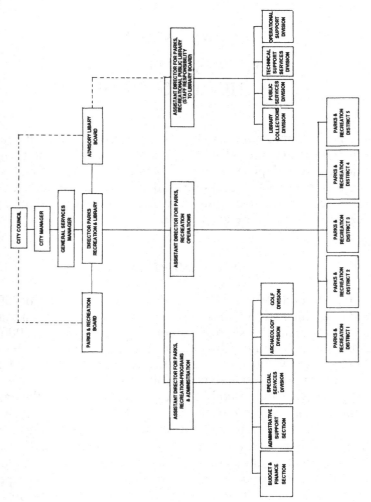

TABLE 4-2

brary administrator could capitalize on the situation to improve considerably the place of the library in the community's governmental-educational-economic fabric.

CASE STUDY

ULISYS IN PHOENIX: THE REMOTE HOST

The Phoenix Public Library system is organized around a central library that provides a full range of services including research and in-depth reference capabilities. It also provides centralized materials processing and collection support for nine branches and one bookmobile, which are geared to popular reading and limited reference service.

ANCIENT HISTORY

By the mid-1970s, the library's collection exceeded 259,000 separate titles and 1.1 million items. Annual circulation reached 3 million, and the service community numbered 675,000 persons. Phoenix and its library were growing rapidly. In 1976, the library underwent several significant organizational and functional changes. First, the central library moved into a fine four-story addition and remodeled the existing structure, thus almost doubling the total size of the facility. At the same time, the collection and staff were reorganized into five new subject departments. This redepartmentalization greatly altered public service patterns, budget allocations, collection development and technical service procedures. Better methods of measuring performance and usage were necessary to manage the new organization. Also, the larger building and divided collection prompted staff concern about remote bibliographic access to the collection since some of the new departments were now at a good distance from the public card catalog, or were on a different floor completely.

In addition to the changes at the central library, construction of the new Cholla branch in a fast-growing, affluent area of the city neared completion. Circulation at the new branch was expected to become the highest among the branches, which were all experiencing a steady increase in circulation at rates equal to or greater than that of the central library. The library's Recordak circulation system, however, simply could not keep up with the increased activity. During fiscal year 1975-76, the library's overdues section came under the scrutiny of the City Council. A large backlog of unprocessed fine notices had accumulated, and retrieval of overdue books had fallen accordingly. In response to the Council's concern, the library proposed that the circulation system be automated. The automated system was expected to (1) reduce the backlog of fine notices and recover additional overdue items, (2) process future fine notices in a more timely and efficient manner and (3) monitor the circulation at all agencies simultaneously as a way of preventing patrons from going branch to branch and abusing borrowing privileges.

The conditions of growth, along with the need to streamline the overdues function, indicated that the time was appropriate for automation. In April, 1976, the City Council authorized the Management Information Systems (MIS) department, in cooperation with the library, to investigate the relative feasibility of either developing or purchasing a computerized, online circulation system. Three MIS computer systems analysts were joined in the project by the library director and the technical services administrator (TSA).

Even at this early point it was apparent that MIS was to take the lead in the automation project. The library administration devised no strategy either to take charge themselves or to increase their role in the decision-making and selection processes. No new personnel with automation experience were sought, and no survey to assess resident expertise among the staff was conducted. No real consideration was given to the possibility of forming a separate automation section within the library. And because city policy at that time stipulated that MIS would operate all the city's computers, the in-house configuration was never seriously entertained as an option.

Instead, it was decided by the library administration that the circulation and overdues functions would continue to be the responsibility of the TSA after automation, and that she, therefore, would represent the library in its dealing with MIS and be responsible, from the library's side of the project at least, for system maintenance and for research and development.

Of course, the year was 1975-76, which was still early in the era of library automation. Many of the options and procedures that today seem obvious were then known only to a few. Computer expertise of any degree among librarians was restricted literally to a handful, and available systems were few in number and limited mainly to circulation control and acquisition functions. Online catalogs were a topic for futurists, and the "integrated system" was as yet unchristened.

Yet some systems were online and some expertise, however scarce, was available within the profession. Although the Phoenix Public administration did some research on the subject of circulation systems, a little more thought on library automation in general would have been beneficial. In-depth consultation with established experts would certainly not have substituted for any lack of computer expertise or experience, but it may well have supplied the library administrators with needed preparation for the crucial, upcoming interaction with MIS and with the system vendors. Instead, many of the library's actions were determined more by default than by conscientious and well-informed decision making.

This lack of involvement may have come about through a lack of awareness, a fear of computers or a professional aloofness from the mundane details of circulation and overdues files. A likely contributing factor, in addition, was an unwillingness to engage in interagency cooperation with MIS, for the project to automate the circulation system was not the first encounter between the two agencies. In 1975, the library purchased an automated BATAB batch acquisition system which again was the nominal responsibility of the TSA, but which in reality, again according to city policy, was controlled by MIS. Personnel at MIS were slow to respond to the library's wishes concerning the system and were reluctant even to deal with the vendor, preferring to handle the system themselves. As a result, system enhancements and upgrades were repeatedly neglected, and the system did not perform at peak efficiency. The lack of agreement between the two agencies and the library's failure to keep the system current and to be assertive in developing the system according to the library's needs were samples of the difficulties to come when the project to automate the circulation system was undertaken.

The circulation project got underway when one of the MIS analysts and the TSA were assigned responsibility for evaluating the options of either purchasing a turnkey system or developing a system locally. They first examined the feasibility of developing a local system to run on the existing computers at MIS. With information gained here as a basis for comparison, they then traveled to Las Vegas, Nevada, to observe the CLSI installation at Clark County Public Library as an example of a turnkey system.

The results of the preliminary evaluation were reported to administrators from MIS, the library, and the city manager's office, who dismissed the possibility of developing a system locally for three reasons: (1) such a system would cost more to develop and would result in higher annual operating costs; (2) the extra load of the 26 terminals projected for the library's system would overburden existing hardware; and (3) local development would take too long (Cholla branch was scheduled to open within months, and no budgetary provision had been made for the purchase of a Recordak system for installation in the new branch).

These decisions seem logical, but they are curious for a number of reasons. First, the initial study ignored another alternative to turnkey purchase or local development, namely, that of purchasing a system developed at some other library and adapting it to the MIS hardware. Again, this oversight may have resulted from insufficient research into the available options. In any case, based on the subsequent behavior of some MIS personnel and on their wish to maintain the software themselves, as discussed below, this might have been the most attractive arrangement. Second, the lack of a Recordak system, itself a relatively inexpensive item, hardly seems a pressing enough issue to justify a large automation decision. If necessary, back-up machines from other branches could have been temporarily installed at Cholla to allow more careful evaluation of the options and of library automation in general, or even development of the system locally. Third, purchase of another computer to host the library's system would have avoided any overloading of the existing equipment. And since the hardware purchase was also necessary for the turnkey option, using this necessity as a justification for eliminating the local development option is groundless.

Nevertheless, the reasons mentioned above were given to the library as justification for proceeding with the turnkey option. This in itself indicates (1) the scope of the communication problems between MIS and the library, and (2) the library's lack of assertiveness and awareness of the options that were open.

Once the turnkey option was chosen, the team of five began the specification process. MIS took the lead throughout. There was no bid conference, no use of outside consultants, and no sending of a draft version of the specifications to the vendors for their preliminary comments, any of which would no doubt have better acquainted the investigators with the nature of the library automation market. When the specifications were completed, MIS sent the Request For Proposal (RFP) to eleven firms that seemed likely to have systems of the type desired. The firms receiving the RFP were

> CLSI
> 3M
> Gaylord
> Honeywell
> DEC
> Control Data Corporation
> NCR
> Univac
> Burroughs
> IBM

Universal Library Systems, Ltd., the parent company of the ULISYS automated circulation system which Phoenix Public eventually purchased, was not even an original recipient of the RFP. DEC, which did not have a circulation system on the market, forwarded the RFP to Universal because the ULISYS system is based on DEC equipment.

Formal proposals were received from

NCR
Gaylord
CLSI
3M
Universal

On July 21 and 22, 1976, MIS personnel, with no input from the library, evaluated each proposal on ten attributes, variously weighted. The following instructions applied to the scoring:

Costs

Scoring should be based on one vendor cost against another.

Capacity Now

The system must be capable of serving the main library and its nine branches with CRTs and light pens. Twenty-six terminals are required. Scoring will be either 0 or 10, depending on having capacity or not.

Capacity Future

The system must be capable of growing to fit the needs of an expanded community through additional branch locations and/or terminals. Scoring should reflect growth potential.

Compliance to Bid

Does vendor comply fully, partially, etc? Score accordingly.

System Software Features

What conversion, application and operating system software does vendor supply? The minimum per bid specification or something exceeding the minimum? Bid minimum is worth 5 points. Any other score reflects + or − features. Simplicity vs. complexity is desirable.

System Software Maintenance

How is the application and operating system maintained? Who maintains it? Is local maintenance provided? Score accordingly.

Application Performance History

Is proposed application running elsewhere? How many places? How large is the library? Are users satisfied? Score accordingly.

Company Stability

How old is the company? What is the company's financial solvency? What is their growth picture? Score accordingly.

ULISYS had the lowest bid and received the highest score, 25% higher than the next closest competitor. Although very expensive, the 3M system, which was adapted from a system developed in Arlington, Virginia, was the runner-up. These two systems, along with CLSI, remained in contention until demonstrations were arranged a few weeks later.

NCR received no score as insufficient data was supplied with the proposal. Later, the evaluators learned that the system did not have online capabilities, which was not in accordance with the bid specifications.

Gaylord was likewise rejected because the system used a non-local host computer in Syracuse, New York. The RFP specified that the computer must be installed locally. Gaylord personnel in fact objected to the RFP on the issue of computer location. They felt that it had been written with CLSI in mind.

CLSI itself, however, was later rejected for three reasons: (1) the hardware configuration used equipment of numerous vendors, with unique operating system software; (2) the system's capacity, at that time, was limited to 16 terminals, while the Phoenix installation required at least 26; and (3) the firm had just passed through a period of financial and corporate instability.

As with the feasibility study, the reasoning upon which MIS made its final decision is puzzling. In the first place, ULISYS, like CLSI, also came to depend on an array of hardware produced and maintained by several vendors, which eventually created severe system difficulties, as will be discussed later. Also, CLSI soon developed an expanded capacity that would have been suitable to the needs of Phoenix Public. Furthermore, neither 3M nor Universal had yet installed their system in a major library. ULISYS was adapted from a system developed and used at the University of Winnipeg, but this was a prototype system with only three terminals. The 3M system, on the other hand, appeared to have a superb corporate and technical pedigree. Nevertheless, it was not favored. This, of course, was fortunate since the 3M system was withdrawn from the market shortly thereafter. ULISYS was chosen even though Universal, at the time, had large financial liabilities.

Thus such things as company stability, hardware array, performance history and future expandability were evaluated with some inconsistency. More important in the system selection process were cost and bid compliance, but also important were capacity and system software. These last are particularly telling in retrospect because CLSI was faulted for the uniqueness of its operating system, which presumably was unfamiliar to MIS, and for the inability of CLSI to supply the needed number of access ports, although a slight extension of time by the city evidently would have allowed CLSI to meet this criterion. Instead, in the end, MIS chose ULISYS, which not only had the needed terminal access capacity but also a more familiar operating system.

The motive of MIS in the maneuvering seems clear. As described below, after the ULISYS installation MIS delayed signing Universal's maintenance contract despite Universal's repeated urging to do so, and insisted on performing system software maintenance itself, much to the chagrin of the library. From the very beginning, there was apparently no intention among MIS personnel, and possibly among the city administrators, to allow the vendor to maintain its own system. Hence the desire for a familiar operating system.

Perhaps MIS did not believe Universal capable of maintaining its ULISYS system. This, however, not only throws the entire bid specification process into question, but hardly seems realistic. Almost certainly the prospect of greater economy was a factor in the decision of MIS personnel to maintain the system locally. In any case, the situation was a recurrence of the relationship between MIS and BATAB, and another demonstration of the library's inability to control the situation even after the BATAB experience.

After a ten-year interim, it is easy to criticize, perhaps too easy. The Phoenix Public Library, though powerless in the affair, was farsighted enough to automate very early in comparison with other libraries. Moreover, MIS made some excellent decisions: 3M failed, but ULISYS survived and attained the major goals the library set for it. In 1976, when automated library systems of any type were largely an unknown quantity, the library's lack of decisiveness is understandable, and the strategy of MIS personnel to bank on their own expertise is not as unreasonable as it might now seem.

The details of the evaluation by MIS of CLSI, 3M, and ULISYS follow:[7]

SYSTEM HISTORY

CLSI — Has had systems in the field for three years. Presently operating in more than 50 libraries. For the most part, users are pleased with the CLSI product.

3M — Has no systems in the field. Equipment for the first system (Princeton University) will be shipped within weeks.

ULISYS — Has one system in the field,[8] has been operational for almost a year. The software package has been developed for marketing by ULISYS. The system was in development for four years at the University of Winnipeg.

SYSTEM APPLICATION FUNCTIONS

CLSI — Transaction functions satisfactory. All necessary reports are provided.

3M — Transaction functions satisfactory. Some could not be demonstrated because of problems in those program modules. All but two reports are available; those can be developed.

ULISYS — Transaction functions satisfactory. All necessary reports are provided.

SYSTEM HARDWARE

CLSI — DEC PDP 11/05 with 56 KB of memory.

3M — Data General Nova with 96 KB of memory.

ULISYS — DEC PDP 11/70 with 256 KB of memory.

SYSTEM OPERATING SYSTEM

CLSI — Unique to CLSI — could not be used for other applications. Programs written in CLSI's FLIRT language.

3M — Data General's RDOS — could be used for other applications but no provision for user education; also core capacity could be restrictive. Programs written in FORTRAN.

ULISYS — DEC's RSTS, could be used for other applications. Provisions for three weeks user training at DEC education centers. Core capacity would allow additional uses. Programs written in BASIC Plus.

SYSTEM PERIPHERALS

CRT

CLSI — Satisfactory operation, large screen with easy to view displays.

3M — Satisfactory operation, small screen with easy to view displays. Space requirements are minimal.

ULISYS — Satisfactory operation, large screen with easy to view displays. Large space requirements. Has light pen attached. Cost to add additional CRTs is smallest of the three.

LIGHT PEN/CHARGER TERMINAL

CLSI — Monarch — satisfactory, uses prompting lights.

3M — Identicon — satisfactory, more bulky and harder to hold than Monarch. Has display capability that CLSI and ULISYS do not have.

ULISYS — Monarch — satisfactory. The pen is attached to the CRT, has no display or prompting of its own.

DISK AND TAPE UNITS

CLSI — 33 MB capacity disk, 4 maximum, standard tape unit.

3M — 33 MB capacity disk, 4 maximum, standard tape unit.

ULISYS — 100 MB capacity disk, 8 maximum, standard tape unit.

PRINTER

CLSI — 340 lines per minute.

3M — 340 lines per minute.

ULISYS — 300 lines per minute.

BACKUP UNITS

CLSI — None.

3M — One per location of each terminal controller. Allows 15 minutes of transaction recording without power.

ULISYS — Portable unit available for $9,000 that could be used for backup or in the bookmobile.

HARDWARE MAINTENANCE

CLSI — Uses CLSI personnel based on the coast.

3M — Not final yet; would probably use a local firm providing service for other 3M products.

ULISYS — Uses local DEC maintenance personnel.

SOFTWARE MAINTENANCE

CLSI — Uses the trouble desk concept with toll-free line. Any problems not handled by phone would require out-of-town systems assistance.

3M — Uses the trouble desk concept with toll-free line. Any problems not handled by phone would require out-of-town systems assistance.

ULISYS — Uses trouble desk concept, but ULISYS will have remote terminal access to the system for diagnostic and corrective action.

COMPANY FINANCIAL STATUS

CLSI — Recent recall of outstanding loans and resale of the company to CLSI's management personnel. From all indications the company is on a better basis than before.

3M — Excellent, with extensive corporate resources to back this new product venture.

ULISYS — Young company with large liability picture. Extended outlook should be good if library circulation control system marketing is successful.

PERFORMANCE BOND

CLSI — Yes.
3M — Yes.
ULISYS — Yes.

MIS was responsible for the scoring. No library personnel participated. When CLSI officials received the evaluation results, the company president and the sales representative requested and were granted a meeting with MIS personnel, library administrators and officials from the city manager's office. In the meeting, the CLSI representatives maintained that their system could handle the necessary number of terminals by linking two minicomputers together. As they had never attempted the procedure, however, no one knew for sure if it would work.

MIS was unconvinced, partly because of some research they carried out on their own. During the evaluation, MIS in Phoenix telephoned their counterparts in Tucson and learned that the CLSI equipment slated for installation in the Tucson Public Library had been sitting idle for months, while Tucson and CLSI wrangled about whose job it was to bring the system up. MIS decided that if CLSI could not accommodate the configuration needed by Tucson Public, it certainly could not handle the more complex arrangement proposed for Phoenix Public.

After the evaluation, all that was left were the system demonstrations to confirm the evaluations. Both the top scoring system and the runner-up were observed at this time. CLSI had been previously observed in Nevada. ULISYS personnel came from Canada to DEC headquarters in Los Angeles and installed ULISYS on a PDP 11/70. On August 4, they gave a mock-up demonstration using three terminals to the same two observers from Phoenix who had visited the CLSI installation in Nevada. On August 5, the observers traveled to St. Paul for a demonstration of the 3M system at company headquarters. Unlike ULISYS, 3M could not demonstrate all system functions. Several were still under development.

The demonstrations confirmed the evaluations, and the contract for the automated circulation system went to Universal. Phoenix Public thus became the first library to purchase ULISYS. The cost for hardware and software was approximately $345,000. Another $30,000 went for labels, site preparation, and other initial expenses.

CONVERSION OF THE STAFF
AND COLLECTION

The first terminals arrived in January 1977. Conversion and training began almost immediately. Four terminals were installed temporarily in the basement of the library, one on the second floor in technical services, and two at MIS. ULISYS personnel installed on the system a test program of limited storage capacity for training exercises. A small number of books were repeatedly converted and deleted to familiarize trainees with the system functions during the week-long training period.

The ULISYS representative trained the TSA, the head cataloger, and six CETA temporary employees, who would do most of the conversion. The head cataloger in turn trained the staff in cataloging services. After that, all new books were converted as part of the cataloging process. The TSA trained the circulation, overdues, and branch staffs. She visited each branch twice, once for training prior to conversion, and again a few weeks later before the facility went online.

To convert the central library collection, staff pulled books from the shelves and attached a barcode label to each pocket. The labels supplied by ULISYS for the conversion process included a detachable strip with a matching number for each label. These strips were attached to the book cards, which were collected, photocopied ten at a time, and quickly placed back in the books so that no time was lost in returning the books to the shelves.

The conversion process proceeded from the photocopies of the book cards after these were edited to delete extraneous information. Using this system, the library was able to remain open during the conversion and circulate its books as usual. No facility went online until the entire collection at that facility was converted. Circulating books were converted first; books returning to the library from circulation were converted before going back on the shelves. There was no parallel running of the automated and the Recordak systems except at the branches.

Branch conversion priority went to Cholla. After that, branches were converted in pairs, beginning with those having the highest circulation. After one pair of branches were converted and online, conversion of the next pair began. Except for Cholla, branch conversion proceeded directly from books wheeled to the terminals on trucks. Cholla book cards were pulled and sent to the central library to be photocopied and entered because no branch terminals had arrived by that time.

Originally ULISYS was a short record system. During the conversion, bibliographic data input included author's last name, full title, material type, branch, location (i.e., reference, circulation, bindery, etc.), level (adult, youth, juvenile), call number and price. In 1982, however, a major upgrade provided the capability of inputting a full bibliographic description with the searching enhancements that virtually made the ULISYS circulation system an online catalog. The early short records required reconversion, however, before they could be searched using the enhancements.

The major weakness of the system was its lack of bibliographic authority control. This greatly increased the possibility of error in the book entry process since essentially every copy of every title had to be entered as a unique record. Misspellings were accepted by the program, with the result that even slight variations in author or title data caused non-linking with duplicates already in the bibliographic database. Hence, in a title search, which was one of the few searches then available, multiple options could appear for any single title of which the library owned multiple copies.

It also became apparent that the CETA personnel should have received more training and supervision during the conversion. Many items were input improperly, misspellings were common, and many titles were entered under initial articles. Hiring untrained workers and allowing them to construct the database proved to be a dubious strategy with residual effects that would come back to haunt the library.

The end result of the above two conditions was a database that lacked sufficient quality to permit further enhancements and more sophisticated automation applications in later years.

Cholla opened in May 1977, fully online. The central library went online in July. The last branch went online in February 1978. Twenty-six terminals were installed with varying degrees of site preparation necessary, mostly drilling for telecommunications lines and relocation of check-out desks. Cholla and the remodeled central library were planned with automation in mind, and so little modification was necessary after the hardware arrived. The computer itself went, of course, to the city's computer center.

The bookmobile was the only facility not to come online even though the books were converted. MIS purchased a portable light pen unit with limited data storage capacity to interface with the computer via modem. The intention was to dump stored data periodically into ULISYS. Unfortunately, MIS made the purchase without consulting Universal, and then neither MIS nor Universal could make the unit compatible. Universal did in the end develop a workable interface, but MIS was reluctant to make use of it. After six years, the bookmobile was still the only unit using a Recordak system.

This was one of several instances in which the library was caught in the middle. It could not make headway with MIS; nor could it go to Universal directly to solve the problems as MIS was the city's bargaining agent in the matter.

In an overall evaluation of the project, however, ULISYS proved to be a resounding success in terms of the original goals of the automation project. The backlog of fine notices was quickly processed, and the fines function operated much more efficiently after installation. In fact, the staff in the overdues section was cut from six in 1976 to two in 1982, and later to one. Furthermore, book reinstatements jumped by 132.5% in a single year. And the Cholla branch, although it opened late, opened fully online.

In terms of system management, however, the project was less satisfactory. The database that was created was severely lacking in quality, rendering it unusable later as the basis for an online catalog. Of equal importance, the library was unable to make optimum use of the available computer expertise at MIS. Instead, the library and MIS were often at odds with each other during the early phases of the project. As a result, proper utilization and development of the system were hampered until the library was able to establish an improved relationship with MIS and other agencies of the municipal government.

INTERAGENCY DIFFICULTIES

MIS almost singlehandedly decided what was best for the library with regard to automating the circulation function. In this respect, MIS deserves great credit for choosing the ULISYS system and for the undeniable benefits the library derived from the system. Left to its own resources, the library could have done worse. But the independent attitude of some MIS personnel led to friction with the library and with Universal. Two examples are disagreements that arose over copy numbers for books and over the software contract.

During the installation, the TSA repeatedly urged the inclusion of a field in each bibliographic record for copy numbers. Otherwise, in the case of multiple copies, it would be difficult and perhaps impossible to determine which copy should be marked off the shelf list when ULISYS reported a book lost. In other words, there was no way to correlate a book's barcode number, by which it was filed in the computer, to its copy number on the shelf list. MIS personnel could not follow that logic. Since each item had a unique barcode number, they felt copy numbers to be unnecessary. They could see no reason to link ULISYS to the shelf list.

MIS would have been correct had the library chosen to discard the shelf list and use ULISYS as an online shelf list instead. But the ULISYS database, as mentioned earlier, lacked the necessary integrity for such a duty. Furthermore, searches by call number were not possible on ULISYS at that time. Technical services, therefore, still needed the shelf list, which in turn necessitated some kind of correlation between the barcode numbers and the copy numbers.

The Universal programmer creating the library's files was neutral in the squabble. Since this was the first ULISYS installation, neither he nor his company had experience enough to make a contribution. In the end, the field for copy numbers was not included.

When ULISYS came online, shelf list difficulties occurred just as predicted. ULISYS reported lost book barcode numbers as well as titles and branch locations for those items, but as it could not furnish enough information to distinguish the lost item from duplicates at the same facility as listed on the shelf list, the item could not be marked on the shelf list as lost. Clerks were instructed to write the barcode numbers of the lost books on the backs of the appropriate shelf list cards until discarding or transfer of the other copies hopefully identified the missing copy by process of elimination. This laborious procedure made collection development difficult and increased the clerical workload in technical services.

As a direct result of the lessons learned at Phoenix Public, Universal provided a field for copy numbers in the software of subsequent installations. But it was not until the 1981 upgrade that the program at Phoenix Public was expanded to include that enhancement.

The disagreement over the software maintenance contract arose because, as discussed earlier, MIS seemed to feel it could provide better maintenance in house. As a result, MIS maintained the software, while paying Universal $150 per month for consulting services. Library personnel, on the other hand, felt that Universal should provide the maintenance, based on discussions with key Universal officials. Nevertheless, a software maintenance contract was not signed until May, 1982.

A number of conditions may have contributed to the decision by MIS to sign the contract. In 1979, Universal began urging MIS more strongly to sign after becoming aware of a particularly bad spate of down time during which the system was down for 86 hours within a five-week period. The head cataloger telephoned the information to Universal's president hoping to bring about a change in the working relationship between MIS and Universal.

In doing this, the head cataloger was overstepping his bounds with respect to MIS and the TSA, who by then had become completely frustrated by her fruitless attempts to work with MIS and had virtually ceased communicating with MIS personnel. The head cataloger had unofficially taken over many research and development duties and was in frequent, though unauthorized, contact with ULISYS personnel in Canada. He hoped to increase pressure on MIS from Universal to sign a maintenance contract. At the same time, system problems exerted their own type of pressure on MIS to sign.

A serious problem developed in the linkages between bibliographic data and item specific data. In the early version of the ULISYS software, bibliographic data for any title was linked to one of the item specific entries, and then all the item specific entries were linked together. Unlike CLSI, there was no separate bibliographic file as such. If the item linked to the bibliographic information was deleted, the system was programmed to transfer the bibliographic data to the next item specific record in the link. With increasing frequency, however, the links were breaking down, and the transfer of bibliographic data for lost books was not taking place. Bibliographic data for books in hand often could not be found in the database. When their barcode labels were scanned, ULISYS would respond BOOK DETAILS NOT ON FILE or BOOK NOT ADDED.

Similar problems developed in the patron files. Patron barcode numbers often pointed to the wrong patron data, and occasionally several numbers linked to one patron.

In addition to these problems, several program enhancements from Universal were implemented by MIS during the first four years of ULISYS operation, such as a program to monitor a rental collection, additions of new material types, and the installation of an interlibrary loan function. In addition, the system's hardware was upgraded several times, including an increase in the number of terminal and access ports.

In the absence of a software maintenance contract, MIS, aside from using Universal's consulting services, had to handle all of these situations alone. The frustrations must have made a software maintenance contract very attractive. An apocryphal tale suggests that when the loudspeaker at MIS announced another ULISYS malfunction, programmers scattered like marbles.

Finally in November, 1981, with a massive ULISYS upgrade underway, MIS endorsed Universal's recommendation that Phoenix purchase software maintenance from Universal. This came in response to a letter from the president of Universal in which it was pointed out that over the previous five years MIS had spent about 18 person-months per year, or a total of $180,000, to maintain a software package originally costing $46,000. The contract was signed in May 1982, much to the relief of the library.

Once the maintenance contract was signed, Universal began providing software maintenance for $1,000 per month. For this fee, Universal provided

- a designated terminal in their service center to receive routine messages, questions and problem reports, which were reviewed and answered daily by Universal personnel,
- a 24-hour, seven-day-per-week answering service to relay important messages to Universal staff members on call to remedy system problems,
- a direct-dial number for immediate action on critical problems affecting day-to-day library operation,
- a minimum of two site visits per year by a Universal representative, averaging approximately five days per visit.

In addition, MIS was still free to write, change, modify or delete any programs, using its own personnel or Universal staff. Also, new software upgrades were supplied at no charge.

RECENT HISTORY

In March, 1980, after a previous telephone discussion with the library, Universal announced publicly that a major ULISYS software enhancement was available that would permit database searching by author, title, author-title, subject, ISBN, LC card number, and call number.[9] No immediate action was taken by Phoenix Public toward obtaining the upgrade because the library was in the middle of an enormous reorganization project. One of the results of that reorganization was the selection of a new technical services administrator.

The new TSA was chosen from the technical services department. He had worked extensively on the ULISYS project, and was, therefore, familiar with both MIS and Universal. This person was soon able to effect an arrangement between all parties for acquisition of the upgrade. MIS and the library agreed to divide equally the cost of $7,500 for a ULISYS programmer to install the new software, plus an additional $2,000 for hardware enhancements. The software enhancements themselves were free according to Universal's software policy, which was extended to Phoenix Public, even though at that point no software maintenance contract had been signed.

It should be mentioned that both the acquisition of the upgrade and the signing of the maintenance contract came about through the efforts of the new TSA. Having established a good personal relationship with personnel at MIS as well as at Universal, he was able to pinpoint program blockages, in whatever agency, and take steps using a combination of persuasion, logic, persistence and forcefulness to maintain forward motion between all three parties. Although the new TSA worked in general through proper channels, it must be added that he was at the same time familiar with backdoor approaches and used them effectively.

The upgrade process began in Phoenix on April 21-23, 1981, with a series of conferences between Universal, MIS, and library personnel. In the meetings, the library presented a list of problems the staff had encountered with the system over the previous five years, along with a list of desired enhancements. The problems centered about four areas: garbage in the database, lack of sophistication in many functions, down time and poor response time, and inadequate statistics. Universal responded by explaining which enhancements could be provided and which problems could be solved in the upgrade. Some of the problems were beyond the scope of the upgrade because they were caused by human error on the part of library or MIS staff. Universal agreed to correct the system errors through the upgrade, and purge much of the garbage from the files to allow the library staff to re-enter proper data. Universal also made recommendations to the library and MIS about correcting human and procedural errors.

A ULISYS programmer came to Phoenix and installed the upgrade late in 1981. The enriched ULISYS came online early in 1982, after staff had received their training. The ULISYS representative led two training sessions for branch personnel, and the TSA trained the central library staff shortly thereafter.

The upgrade made ULISYS a different system from the one purchased in 1976. Mastering it was in a way like starting all over again. For instance, the system now had a separate bibliographic file that would accept a full bibliographic record and serve as an authority for holdings records containing only item specific details. Linking these two files together minimized the possibility of error when entering duplicates and ensured higher quality in the database. It also, however, increased considerably the amount of data that had to be entered for each title.

ULISYS now had the authority control and searching features of an online catalog. But this required a database of full bibliographic records, and Phoenix was burdened with a database of short records and insufficient personnel to enrich the old records — almost 350,000 of them. Several options were presented. First, the database could be allowed to remain "split," containing both full records for books added after the upgrade and the short records input previously. This, however, would preclude searching of the 350,000 old records. Second, a partial retrospective conversion could be undertaken, manually enriching records for items selected by year of publication, popularity, etc. Third, extra personnel could be hired temporarily to do a complete manual conversion. The experience of using untrained workers during the 1976 conversion made this option unattractive, however. Fourth, the old records could be matched by computer against the database of a vendor, a bibliographic utility or another library with a collection profile similar to that of Phoenix Public. The full records for all matches could then be copied from the host database and loaded into the new ULISYS files, thus minimizing the number of records needing manual conversion. This would require, unfortunately, a massive clean-up of the current database to eliminate garbage and to merge all multiple entries into a single, accurate record for each unique title.

The final option was selected for upgrading the bibliographic database. As a first step, library staff were assigned the duty of examining printouts of all the records in the database, correcting the errors and merging variant entries. At the completion of that lengthy process, the plans call for dumping the records to tape and matching the tape against the database of a library that has agreed to assist in the conversion process. The records obtained from the matches would then become the bulk of the new ULISYS database.

The process of creating an online ULISYS catalog was facilitated by the fact that Phoenix Public finally joined OCLC in 1981. It was an event that was painfully overdue, given the size and nature of the collection at Phoenix Public. Nevertheless, the library was never able to convince the city budget makers of the advisability of joining a bibliographic utility. Only after the new TSA and a new library director located a source of alternative funding did OCLC membership become a reality for Phoenix Public.

THE FUTURE

The future state of ULISYS in Phoenix is complicated by two factors. First, the technical services administrator who was instrumental in achieving a successful working relationship between all three parties involved in the ULISYS project left the position after less than a year. Since then, two others have held the position in quick succession. Such administrative instability in the key library position for the ULISYS project threatens to undermine progress made up to now.

Second, the ULISYS system in Phoenix has been plagued by severe hardware difficulties since the upgrade. Between December 1982 and July 1983 alone, hardware failure resulted in three separate down periods. The worst instance began on May 31, when a disk drive crashed. Next it was the tape drives. After a number of equipment replacements and subsequent failures, the system finally began to stabilize as of July 15. Yet even then the holds function remained inoperable, and the fines function was only intermittently operational. In just three days during this disaster (July 1, 2, and 5), the records for 43,280 check-outs and 49,358 check-ins, or a total of 92,638 transactions, were lost. At one point, MIS personnel feared that the entire ULISYS database had been irretrievably lost.

Even though the problems began with the hardware, the crash involved a combination of hardware, software and human failure. The situation was greatly complicated by the fact that the hardware was supplied and maintained by many separate vendors. Coordinating their efforts during the emergency was very difficult. One of the vendors was unable to locate replacement parts anywhere in the country. During the early days of the summer crash, even the exact nature of the problem was a mystery. Entire files had to be rebuilt from the backup tapes, and a programmer was dispatched from Universal to help with the project. Then it was discovered that some of the tapes were faulty as a result of a tape drive malfunction that had occurred during their creation.

Relations between MIS, Universal and the hardware vendors became strained during the crisis, as each wanted to place the blame for the crash on the others. During much of this time, the library was not directly involved in the decision making, although MIS, Universal and the library communicated freely and openly throughout. Because of the generally improved working relationship established between them a year earlier during the upgrade, progress was not halted, and the problems with the system were properly addressed.

There is little doubt either that as the library was part of the much larger Parks, Recreation and Library Department, the city government gave more attention to the library's difficulties. After the crash, the library presented its arguments for the design and acquisition of a completely new automated system before 1987. The new system would incorporate the latest technology and provide the fail-safe backup systems and redundancy that the current system lacked. Also, the host computer would be located in the library.

Surprisingly, the same city government that repeatedly rejected the library's proposal to join OCLC readily agreed to begin the planning phase for a new automated circulation system. In October, 1983, a library staff committee was formed to develop general goals for the new system and propose functions deemed by the staff to be essential or desirable in the proposed software. Later that year, using the committee's recommendations, the head cataloger and an MIS analyst began a series of intensive projects to plan the requirements for the new system. These plans, drawn up over a three-month period and totaling over 1,000 pages, focused on the nature of library operations and on the specific needs of the Phoenix Public Library. The plans were then used to formulate the 100-page RFP for a fully integrated online system. A prominent consultant was retained to assist in the RFP formulation and bid process, and city officials gave their full support and commitment to the acquisition of the new system.

It is clear that this second automation project was conceived and initiated in a manner much more favorable to the needs of the library. MIS supplied the technical expertise, and the library supplied the application. This change in approach from the first project can be attributed in part to the fact that Phoenix Public now has ten years of automation experience, and in part also to significant developments in the field of library automation since 1976, when Phoenix Public began the ULISYS project. In a larger sense, however, the clarity of purpose apparent in the new project reflects the improved working relationship the library has been able to establish with MIS and, on another level, to the integration of the library into the mainstream of the city government.

T. D. Webb

REFERENCES

1. Malcolm Getz, *Public Libraries: An Economic View* (Baltimore, MD: Johns Hopkins University Press, 1980), p. 173.

2. Getz, p. 172.

3. Getz, p. 173.

4. Lowell Martin, "Emerging Trends in Interlibrary Cooperation," in *Cooperation Between Types of Libraries: The Beginnings of a State Plan for Library Service in Illinois.* 16th Allerton Park Institute, 1968, ed. by Cora E. Thomassen (Urbana, IL: Graduate School of Library Science, 1969), p. 9.

5. Martin, p. 8.

6. For a more complete analysis of the reorganization of Phoenix Public Library, see T. D. Webb, *Reorganization in the Public Library* (Phoenix, AZ: Oryx Press, 1985).

7. These details are from the technical service administrator's official report on the automation project and are, therefore, representative of the selection process as understood by the library.

8. The reference here is uncertain, but presumably refers to the prototype installation at the University of Winnipeg. Phoenix Public was Universal's first Library contract. See "Universal Lib. Systems, Dataphase, Gaylord Win Circulation Contracts," *Advanced Technology/Libraries* April 1977:1–2.

9. "Public Access Terminals for California Campuses," *Advanced Technology/Libraries* March 1980:6.

Chapter 5

The Joseph F. Smith Library:
Introduction to the
Case Study

The case study of the Joseph F. Smith Library has a double focus. First, it details certain operational aspects of an in-house configuration and shows that the library experienced a number of problems remarkably similar to those of Phoenix Public. But because its computer was located in house, the staff of the Smith Library performed much of the corrective action itself. The Smith Library had to provide a high level of technical expertise when dealing with the problems even though maintenance contracts were in effect with the software and hardware vendors from the date of installation. As the case study shows, the corrective actions taken by the Smith Library often brought about no more effective remedy than that experienced at Phoenix Public even though the computer was located in the library.

Secondly, the case study focuses on the Smith Library's plan to utilize its system to fullest capacity, well beyond the limits of the turnkey software. Since maximization of automated resources is a major theme of this book, the case study will provide an account of the project to integrate several locally developed modules with three vendor-supplied modules, thus creating a hybrid system more suited to the specific needs of the Smith Library.

The Smith Library is the major component of the Division of Learning Resources, which is one of seven academic divisions on the Hawaii campus of Brigham Young University (see Tables

5-1 and 5-2). The concept of the learning resources center, in which the library is only one part of an integrated set of media collections and services, is more common at the secondary level and in two-year colleges than in a four-year institution like BYU-Hawaii. In such a facility, the purely bibliographic functions are performed in conjunction with other media storage and delivery services. The Smith Library shares space, budget and staff with an array of non-print functions, some quite sophisticated. The library is closely connected with the university's recording and television production studios, the campus cable television network, the university's satellite earth station, the media center with its media collections and interactive video systems, the media production labs and the instructional technology programs.

The learning resources approach has provided the library with a staff possessing a relatively high level of technical expertise and a proclivity to system development in the areas of technology, electronics and automation. As the case study will show, the Smith Library involves itself in the direct local development of its automated system to a degree many libraries never attain, especially those with turnkey systems. This is possible because of (1) the availability of competent yet inexpensive student computer programmers, (2) resident expertise in electronics and computer technology, (3) an organizational structure that places systems research and development under the supervision of professional librarians, (4) adequate funding from the library's parent organization, and (5) the library's optimum size for automation.

Because the Smith Library is in many ways atypical of libraries as a whole, the case study does not present it as a model that other libraries should emulate in all aspects. Instead, the study is intended to be an account of a particular instance of a stand-alone, in-house automated configuration in which a number of conditions resemble those of the remote host installation at Phoenix Public, despite the obvious and numerous differences. The similarities underlying the differences demand explanation because they help clarify the nature of library automation in general.

The relationship of the Smith Library to its vendor, Dynix, Inc., is also unique and needs to be mentioned. Certain high-

BRIGHAM YOUNG UNIVERSITY–HAWAII CAMPUS

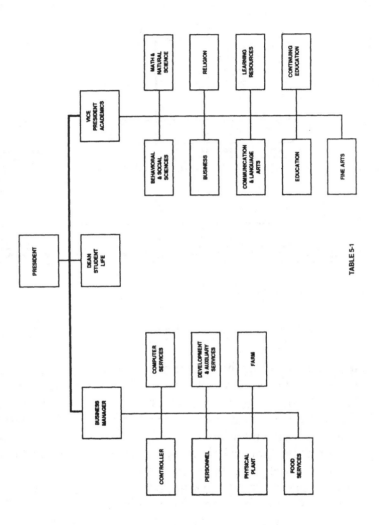

TABLE 5-1

DIVISION OF LEARNING RESOURCES

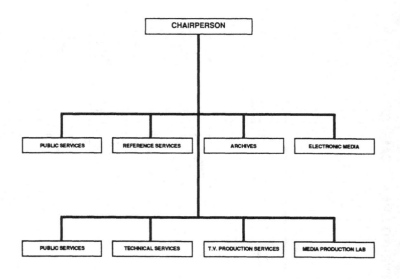

TABLE 5-2

level personnel in the Dynix organization, including one of its founders, are former employees of the Smith Library who gained much of their initial automation experience in that capacity. These were influential in developing the Smith Library's first automated system before taking their skills into the open market. Although this unusual circumstance at times facilitated communication between the library and its vendor, it also was the source of what proved to be some unwarranted and unfulfilled expectations. Each side occasionally expected more favors than the other felt obliged to give based on previous working relationships.

The Smith Library became a Dynix client in 1984. Dynix, Inc., was formed in 1983, and within two years had signed up nearly forty libraries around the country as clients. In doing so, it captured almost 10% of the 1984 new systems market.[1] This spectacular growth has been mostly among small and medium-sized libraries, which constitute Dynix's major market.

The Dynix corporation has received some very good press,[2] and client libraries seem generally satisfied with its system. The Dynix product is very sophisticated although only three modules are available at this writing. These are cataloging, circulation and an online public catalog. Serials and acquisitions modules are under development, and other automated utility programs, such as an OCLC interface and a data conversion package may also be purchased.

The system's sophistication is primarily a result of its architecture. Unlike many "integrated" systems that grew with varying degrees of success from earlier circulation systems, the Dynix system incorporates a truly integrated design by linking all major functions to the central bibliographic data files. Transactions in one module or file effect corresponding data alterations in the appropriate files of other modules. This is accomplished by means of "pointers" that link various files together. Different functions require different files, and the numerous sophisticated functions available in the Dynix software require a proliferation of files to do the work. But the linkages make data duplication within the different files unnecessary, thus saving storage space and processing time.

The Dynix system, then, is sophisticated, but it is a new product. And Dynix, Inc., is a new company. As such it falls into a dubious category described by De Gennaro:

> If . . . one wants to have a system with the latest and most sophisticated design and capabilities, one will have to choose a system that is relatively untested and do business with a new entrepreneurial company that is subject to all the usual problems that beset such companies. Keep in mind that you are entering into a long term dependent relationship with the vendor that sells and supports it. The vendor's troubles become your troubles; the vendor's failures become your failures.[3]

When the Smith Library chose a vendor, it was looking for the "latest and most sophisticated." Dynix had the desired sophistication, and, partly because of its special relationship with the Smith Library, made a very economical bid, even below its usual low market price. The deal was made quickly. But sophistication breeds problems of a corresponding magnitude. The greater the sophistication, the more serious the potential problems. The solution to such problems requires the closest sort of vendor-client cooperation. In this, the Smith Library, despite its in-house configuration, was often no more successful than the Phoenix Public Library. The resident expertise at the Smith Library was a significant factor in minimizing the problems as they developed however. Without that expertise, the in-house configuration would undoubtedly have contributed to a compounding of the problems that were experienced.

CASE STUDY

HAWAIIAN STAND-ALONE

In some ways, Brigham Young University-Hawaii Campus in Laie, Hawaii, is typical of many small, privately owned four-year colleges. BYU-Hawaii can only accommodate a small number of students, not more than 2,000. It offers associate's and bachelor's degrees in a number of fields, mostly liberal arts. The campus and facilities are compact and are situated in a non-urban setting with beautiful surroundings. And avid fans follow the progress of its teams.

In other ways, however, BYU-Hawaii is not typical. For example, it enjoys good financial support from its parent organization, the Mormon Church. Its students come from nearly forty nations, mostly from Polynesia, the Asian rim and the mainland United States. Over the past several years the faculty and students have availed themselves of computer technology to a degree that is surprising in a small college. Fully 70% of the faculty use either mini or microcomputers on a regular basis in support of their professional responsibilities. A standing faculty committee coordinates the burgeoning growth and use of computer resources on campus. There are five computer labs for student and faculty use, and the number of faculty members who own their own computers is growing rapidly.

The university's library, too, is unusual in a number of ways. With 130,000 volumes, it is the largest of the libraries of the state's independent colleges and universities. It is also one of the most fully automated in Hawaii. Each year the library hosts delegations of educators and librarians from around the Pacific and Asian areas who come to learn about library automation and the learning resources center concept. The library recently made an agreement with the People's Republic of China to exchange staff members in order to assist in the modernization of Chinese provincial libraries, particularly with respect to automation.

Despite these unusual circumstances, however, the BYU-Hawaii library faces many of the same difficulties that confront other libraries of varying types and sizes. The case study will show that this library's automation project was no exception.

ANCIENT HISTORY

The Joseph F. Smith Library, like the library profession itself, has evolved through a number of successive phases of automation. Initially, the library spent several years developing a system locally while sharing time on the university's main computer located in the Computer Services department. This original venture into library computerization began primarily as a circulation system. But other functions were gradually added as the need for them was perceived. Eventually, the system included circulation, cataloging, acquisitions, serials and media scheduling, and was in use until 1984 when the Dynix system was purchased.

Most of the programming for this early system was written by librarians on the staff. Most of these were sincere but self-taught computer users who more or less learned as they went. As a result, the development of the system was laborious, slow and costly. The programming often was substandard, as was the overall system design. And the software was infested with all manner of bugs. The media scheduling function, for instance, did not work satisfactorily, and was abandoned soon after its implementation. The COM catalog production program was defective and inefficient. Likewise, the acquisitions program was unfriendly and unreliable, and finally fell into partial disuse.

The circulation system was a success, however, as was the cataloging module. The online catalog was outstanding, and became very popular among library users. It provided author, title, and subject access to the collection, and included a very fine online tutorial. The searching procedures, however, were not very sophisticated, the programming was inefficient by industry standards and the database lacked integrity.

Hardware problems were abundant. In the first place, there were too few terminals available for the staff and students because the number of ports assigned to the library by Computer Services was insufficient to meet the demand. Furthermore, the library system often overflowed its allotted space on the university's computer. By far the largest user on the computer, the library simply did not have enough computer storage space to support its expansive system. Each time the library's disk space was filled, the system locked and all transactions ceased. The Computer Services staff were then notified so that they could delete old records and statistics information or clear unused files in order to provide enough work space for the library system to begin operating again, at least for a time.

Each such incident made the library staff more dismayed with the remote host configuration. Although the relationship between the library and Computer Services was amicable, the library staff felt they lacked real control of their own system and could not develop it further as the need arose. An in-house, stand-alone computer became an increasingly attractive option, one that Computer Services fully endorsed because this would free an enormous amount of storage space and memory on their computer.

While the library contemplated the possibility of acquiring its own hardware, an attempt was also being made to resolve some of the software problems. The library hired a first, and then a second, student programmer enrolled in the university's computer science major to assist in the software maintenance. Initially, it was believed that the majority of the programming would still be written by a librarian and that the students would only be involved with debugging existing programs, making minor enhancements to the software, and performing other specific tasks as assigned and supervised by the librarian in charge of automation. The unexpected expertise of the student programmers, however, eventually secured for them a much more influential role in the library's automation project.

At first, the students had some difficulty learning the library's system. The university's main computer was new to them, and the language was not one of those they had learned in class. In addition, the logic of librarianship was not always easy for them to follow. As is common in many library automation projects, there was not enough common understanding between the librarians and the programmers concerning the system's input and output. The librarians often assumed that the programmers knew more about library procedures than they actually did. Frequently, programming assignments were given to the students without a thorough explanation of the library procedures involved or what objectives were to be achieved.

This condition reflected a general lack of awareness of systems analysis and design procedures and a failure to implement proper systems planning techniques prior to the start of programming. There was a tendency among the librarians to plunge into the coding of the software without adequately assessing the objectives of the project. It is very difficult to reduce even familiar operations to tasks that can be quantitatively addressed by a computer program. This type of analysis requires conceptualizing the operations to be automated in terms of inputs, outputs and parameters. It involves careful and often lengthy feasibility evaluations, system planning and design and perhaps even experimentation with a prototype. Also, regular and intensive discussions with the end-users are essential in order to obtain a clear understanding of what is expected from the system and to get feedback on proposed designs. Likewise, complete communication is necessary among all who will be doing the automating in order to promote the exchange of ideas and the sharing of knowledge.

This planning may be more time-consuming than the actual programming, but it saves time in the long run by reducing error and misapplication of the software. Had the Smith Library performed more careful preliminary analysis and design when developing the original system, the later difficulties and failures mentioned above would have been minimized. Too often, however, the automation librarians assumed they already knew the procedures they were trying to automate and therefore proceeded to define the input and output with insufficient information from the end users.

Another factor that contributed to the lack of consistent planning during this early phase of automation was a marked discontinuity of the professional staff in charge of the automation project. Within a four-year period, four different librarians had charge of the automation department. Each had a different level of expertise and a different perception of what the automation project should accomplish. This was not only true with respect to the individual modules being developed, but also in terms of an overall, carefully planned blueprint for the final system.

General progress was made, however, and after a few years the system neared a state of completion. The student programmers made a significant contribution in the final phases of development of the original system. After grasping the purpose behind each of the modules and gaining a familiarity with the computer and the language used, they were able to take over most of the software maintenance. They also cleaned up much of the programming written earlier by the librarians, making it more efficient and debugging a number of problem areas. Because of their training in computer science and systems design, however, they perceived serious flaws in the basic structure of the software and the execution of the existing programming that prevented any real efficiency in the system's operation. They longed for the opportunity to throw out the old programs and begin again, starting with a carefully planned design for the system.

ENTER THE VENDOR

The opportunity came in the fall of 1984. At that time, the library completely redirected its automation project by reorganizing and restaffing the automation department. These changes were based on a number of considerations, including a desire to overhaul the software and gain control of the hardware, and a recognition of the expertise and professionalism of the student programmers. Their ability, coupled with their relatively low wage rates as students, put the library in a position to dispense with its costly and inefficient software and begin again without incurring prohibitive additional development costs.

The proportions of a new system that would fully automate the library, however, were enormous. Writing and maintaining the software for a new locally developed system would have made undue demands on the relatively small professional staff of librarians, which was already stretched thin. Furthermore, there was an inevitable continuity problem with the students. Eventually they would graduate, and training their replacements to handle such an enormous and complex system did not seem entirely feasible. The library administration decided, therefore, to scale down the amount of local development and integrate this with a suitable software package from a commercial vendor.

Several major steps were taken almost simultaneously. First, the library purchased its own minicomputer and located it in the library. Second, the library purchased circulation, cataloging and online catalog modules from Dynix, Inc., a turnkey vendor located in Provo, Utah. Third, all the existing modules of the library's original locally developed software were discarded. Fourth, the student programmers were retained to assist in the development of new modules to supplement and interface with those available from Dynix.

The new modules to be locally developed included, among others, acquisitions, media scheduling and COM production. As mentioned earlier, these functions had been automated in the original system but were plagued by problems in their structural design, programming and outputs. Theoretically it would have been possible to correct the errors to a certain extent and install the corrected modules in the new in-house system. But the corrective process would have been more lengthy than beginning over again. In addition, these early modules had been written for the university's main computer which, though using a similar operating system, was not of the exact type as that selected by the library. Additional rewriting would have been necessary, therefore, even to make the old programs compatible with the new computer. In light of these considerations, then, it was decided to structure and write a completely new set of modules, and in so doing, to make them emulate the new Dynix formats as much as possible for the convenience of the library staff.

Dynix was selected as a vendor for a number of reasons. First, it provided three basic modules that were very sophisticated in terms of the functions provided. The system architecture was totally integrated and would lend itself well to the library's plans for supplementing it with locally developed modules. Second, the price was by far the most affordable. The total cost of the original Dynix package, including hardware and software, was about $54,000. The yearly software maintenance charges were an additional $2500. The cost of the maintenance contract for the hardware was about $6,000 per year. Third, because a number of Dynix officials were former employees of the Smith Library, an element of mutual trust had already been established.

The computer purchased by the library was an Ultimate 2000S, originally equipped with 256 kilobytes of memory, 154 megabytes of disk storage, and 16 terminal ports. It has since been upgraded, however, and is considered a model 2020 as a result of the installation of a faster processor. The machine now has dual 154 megabyte disk drives, one megabyte of memory, and 32 terminal ports. This upgrade brought the machine to its maximum memory, storage and access levels, and cost the library an additional $20,000, plus an extra $3500 yearly for the enhanced maintenance contract.

INSTALLATION AND TRAINING

The computer was delivered to the library early in September, 1984, by the subcontractor who supplied local hardware maintenance in Hawaii for the Ultimate Corporation. The technicians installed the machine, loaded the system generating software from tape, booted the system, tested their work, then left without giving any instructions about the operation of the machine. They told the library staff that the software vendor, meaning Dynix, would provide any and all training for system operation.

This was not an unusual nor unexpected condition. The relationship between any library and its multiple automation vendors is a legally and technically complicated affair. Hardware vendors, of which there may be several, mind their own business, and so do software vendors. This leaves the library somewhere very much in the middle. In the case of the Smith Library, the staff already had previous automation experience, but on a different type of machine. The experience, therefore, did not automatically make them capable of operating the new Ultimate. Yet it was purchased with the intention of affording the library with increased control of the system. Waiting three weeks for the Dynix team to arrive in order to begin learning how to operate and program the machine was a particularly frustrating prospect.

As a result, the automation staff began to plow through the eight volumes of documentation that came with the machine. This documentation covered such areas as the programming and report languages, file structure, peripheral configurations, and other highly technical areas. The explanations assumed a high level of machine and programming experience. None of the documentation would have been suitable for novices or end users. Much of it was obviously written for experts and was difficult reading even for those of the library automation staff who were accustomed to computer documentation.

Some assistance came from members of the Computer Services department, who were familiar with Ultimate computers. At that time, Computer Services had two Ultimates identical to the one purchased by the library, in addition to the main computer which had hosted the library's original system. This assistance, along with the documentation, enabled the automation staff to begin operating the computer, at least to the level at which simple programs could be written, lines tested, and terminals installed so that everything would be ready when the Dynix personnel arrived for a one-week installation and training period.

Two Dynix representatives arrived near the end of September, 1984. The installation of the Dynix software on the library computer was complicated by the fact that data from the library's original system had to be dumped onto tapes from the university's main computer where it had been hosted, loaded onto the database of the library's Ultimate and reformatted to fit the file structures and record patterns of the Dynix software. This complication, along with other factors, detracted seriously from the effectiveness of the installation visit. The installation itself was delayed, and as a result, the training sessions were less productive than they could have been.

The relationship between a library and its primary automation vendor involves the exchange of enormous amounts of information in both directions. Yet each of the three phases of the library-vendor interaction — i.e., pre-installation, installation, and post-installation — may be characterized by a different predominating type of information flow between the two parties. Each phase also serves as preparation for the next. Failure to establish the proper information flow in each phase not only delays the exchange of information at that point, but also impedes the success of the subsequent phases. In such an instance, progress is hindered as the two parties attempt to gather information that should have been communicated earlier in the process. Once the information cycle is disrupted, the effects can be persistent.

The pre-installation phase is characterized by an information flow from the library to the vendor. The vendor should obtain from the library the institutional characteristics that will serve as parameters and categories for the automated system. Such things include the number and names of different collections, the types of patrons served, circulation features (fine rates, loan periods, etc.), statistical needs and so forth. Exchanging this information early accomplishes two things. First, it gives the vendor a head start on installation, preventing on-site delays and averting potential problems in the setting of the parameters. Second, it makes the library staff start thinking about their habitual tasks in systemic terms. Without this initial exercise, chances for a smooth installation decrease measurably.

Information flows largely in the reverse direction during the installation/training phase. The vendor provides end-user instructions and explains how the parameters already established function to control the system. After all, an automated library system is simply an efficient, mechanical way of imposing a host of restrictive borrowing parameters to a virtually endless stream of transactions.

In the post-installation phase, the information flow is bi-directional. The library tells the vendor about the system's performance and identifies problem areas that need attention. The vendor supplies further training and services as necessary, especially in the form of system enhancements and upgrades.

The Dynix installation at the Smith Library was not optimally successful because the proper information cycles were not established. Many hours that should have been spent in training were spent instead in conference between the library staff and the Dynix representatives identifying and labeling collections, item types, patron types, statistical parameters and other institutional characteristics. Furthermore, the transfer of data from the old system to the new turned out to be more complicated than the Dynix representatives had expected. The data transfer, therefore, took a day longer than anticipated, thus further reducing the amount of time devoted to online training.

The planning for these operations should have been handled before the installation visit. Because it was not, the training was rushed, uneven and incomplete. Recognizing this, the Dynix managers later agreed to provide four days of supplemental training for the library's automation coordinator at the Dynix corporate headquarters to compensate for the training time lost during the installation.

This extra training took place one month after the installation, but was not entirely satisfactory because unaccountably it was not completely transferable to the library's system. For example, the procedures for installing the statistical parameters for tracking circulation by classification division were explained during this visit to Dynix headquarters, because there had been insufficient time for such an explanation during the installation. When the coordinator returned to the Smith Library, however, the new procedures would not work. Dynix personnel were unable to account for the failure, but they discovered that the particular files involved in this operation contained garbage data. The garbage must have been on the tapes the Dynix representatives used for the installation because no one on the library staff knew how to access these files at that time. Yet a Dynix customer service representative insisted that library personnel were responsible, saying that the garbage probably caused the malfunction.

Subsequently, other caches of garbage were discovered. For instance, when members of the automation staff began to define the security configurations for additional computer ports, they found that several ports were already erroneously configured for the terminals of the staff in the Dynix offices in Provo. Evidently, the tape used for the installation at the Smith Library was copied from the Dynix office version without first purging the Provo data from all the files.

Other installation errors also occurred. For example, after the library staff had laboriously delineated the different parameters for each of the library's patron types—students, faculty, staff, guests—all the patron records transferred from the old system were mistakenly assigned a student classification. Since borrowing privileges are controlled by the patron type designation, this error greatly reduced faculty and staff privileges while giving guests unwarranted borrowing latitude.

Similarly, misinformation from the Dynix trainer during the installation led to the mislabeling of virtually all items in the entire bibliographic database with respect to their various collection designations. The library staff was instructed to enter letter codes where descriptive words were required and vice versa. The result was call numbers that were muddled and unrecognizable in the bibliographic screen displays of the search options of all online modules.

These errors required a good deal of corrective reprogramming during the post-installation phase, much of which was done by the library programmers with assistance from Dynix. The most troublesome and persistent installation-related problem, however, involved the program to print overdue notices, which the library programmers were not allowed by Dynix to correct. The Dynix software is designed so that libraries can create free-text messages to serve as headings for overdue, hold, recall and fines notices. The feature was demonstrated during the installation with the creation online of a dummy heading. But after the installation, the function to delete the inappropriate dummy heading and replace it with a suitable message could not be made to operate. The command to delete the dummy heading was entered repeatedly, and the new heading was composed, entered and filed. But the notices program was apparently not linked to those functions, and the dummy heading continued to appear on every notice, rendering them unsuitable for distribution.

Again, Dynix personnel could not account for this malfunction and were unable to remedy the problem in a timely manner. As a result, the library withheld its final payment until a solution to this and other problems appeared. Relations between the library and Dynix began to deteriorate rather badly at this point. But when the notices problem appeared to be resolved, four months after the installation, the payment was released, and both parties attempted to restore any lost mutual trust.

During all this time, the library was unable to mail any notices. The only recourse was to have the automation staff write programming that encumbered student files in the Computer Services database so that students could not register for new classes, graduate or request transcripts until their fines and overdues were cleared at the library. This transfer of data from the library system to the university system was performed by regular data dumps via tape. This process at least secured fines and overdue books for the library, but the inability to mail notices was a frightful disservice to the library's users and resulted in many unpleasant incidents at the circulation desk.

Somehow, unfortunately, solving the problem of the notice headings created another problem in the notice program. Instead of printing the normal 30 or 40 daily notices, the program now began to print an enormous number of meaningless "shadow notices" which resulted in daily print runs of over 600 notices, most of which were invalid. The cost of paper for runs of this size was prohibitive, so once again the library was unable to provide notices for its patrons. And once again, Dynix personnel were unable to account for and correct the problem. Only after seven more months and numerous online attempts were they able to correct the program. In September, 1985, one year after the installation, the library was finally able to produce acceptable notice runs on a regular basis.

Thus the post-installation phase was complicated by problems that originated during the installation and that could have been minimized by more adequate preparation prior to the installation. It is interesting to note that after the Smith Library installation, Dynix instituted a very thorough procedure to gather information from libraries awaiting installation, thus establishing the proper information cycle early and alleviating many of the problems encountered by the Smith Library.

It must be added that the library was not blameless concerning the difficulties that marred the installation. In the first place, the library was not careful enough in specifying ahead of time the training it expected to receive. This should have been included in the contract, but it was not, and added to the confusion and dissatisfaction accompanying the installation process. Secondly, the library staff had not properly prepared themselves attitudinally for the transition from the locally developed system to the turnkey software. The staff was comfortable with and strangely loyal to the existing software despite its many shortcomings. Some were quite reluctant to change to a much more sophisticated system. Third, the Smith Library, like many others in the library community, failed to realize fully that no system, not even a "turnkey," is maintenance-free. Furthermore, it did not fully understand that the potential for, and the complexity of, system problems are directly related to the level of a system's sophistication. Indeed, many of the problems encountered in the Dynix system were caused by the newness and sophistication of the software as compared with the library's older system.

Some of the problems, then, were actually carry-overs from the library's earlier, locally developed system. In addition to those just mentioned, another was the presence of bad data and records in the files that were transferred to the new system. Records that were incomplete or improperly handled as a result of flaws in the old system could not be expected to disappear when transferred to the Dynix files. Earlier errors came back to haunt the library on a regular basis, and had to be corrected as they appeared.

Another more serious carry-over problem involved the barcode labels used at the library prior to the Dynix installation. These labels were locally produced for the library by the Computer Services department on one of their in-house printers. Though adequate for the library's original system, they did not have check digits, separate sequences for patrons and books and other features that are available in commercially produced labels, and that Dynix uses to activate security functions in its software.

Dynix strenuously urged the library to buy new labels and re-barcode the entire collection. Understandably, the library was reluctant to do so. Dynix finally agreed to modify its software so that the system would accept the locally produced barcodes, but warned that grave problems could result. For example, the built-in security functions that were to be keyed by the commercially produced labels had to be disabled.

This may account for the extra delay in the data transfer process during the installation. The technician had to trace each function in the software, find those that used the special features of the recommended labels and remove them so that the system would accept the locally produced barcodes. Furthermore, each time Dynix sent the library a tape with a new release to be loaded, their technicians had to dial in and perform the tedious modifications all over again. With each new release, there was a great risk of an incomplete modification or of inadvertently creating some other type of error.

Another major problem, one that was not a carry-over, but was instead software related, was the fragility of the Dynix public online catalog. Even before the installation team left for home, students using the online catalog were breaking out of the Dynix software into the command level of the Ultimate operating system. This could be accomplished at several places in the catalog by entering garbage or incorrect commands. The Dynix programming at that time simply did not provide adequate input checks.

This constituted a major security problem because, as mentioned before, there were experienced Ultimate users on campus who certainly knew commands that could cause irreparable damage to the library's system if entered at the system level after breaking out of the public catalog. Yet the Dynix customer representatives in Provo reported that the reprogramming to prevent the breakouts would require a major upgrade, which Dynix was then unable to perform. The library programmers, on the other hand, felt that they could incorporate the needed input checks into the system and thus prevent the breakouts. But the Dynix programs were mechanically encrypted and were therefore indecipherable to the library automation staff.

Since the library programmers could not decode the Dynix programs, they could not prevent students from breaking into the command level of the system. Instead, the programmers created a security "shell" around the public catalog so that once at system level, the users were much less likely to do any damage. The programmers accomplished this by redefining all the standard system commands in the shell around the module so that if entered, the commands would take the user back into the online catalog. In place of the redefined commands, the programmers substituted random or nonsense words and defined them as the only system commands that could be used at that point.

Aside from the threat to security, the online catalog fragility was also an annoyance. Once they broke out of the catalog, students did not know how to get back in. Reference workers had to clear the terminals and re-enter the catalog for them. Unfortunately, during peak use periods, this occurred several times per hour and quickly became tiresome. Although some of the problems that allowed this to happen were fixed by Dynix a few months after the installation, others were not resolved until over a year later when a massive new release was installed.

VENDOR RELATIONS

During the first year after installation, a generally good relationship between the library and Dynix was occasionally strained. The Dynix response to problems was prompt, except in the case of the persistent conditions mentioned above. Because the Dynix programs were encrypted, the library programmers were prevented from correcting or even analyzing those problems directly. This was particularly frustrating because the programmers were accustomed to debugging a program as soon as a problem arose. Waiting, sometimes for months, for someone else to fix what the programmers felt they could have corrected quickly themselves was annoying and contributed to the strained relations between the library and Dynix.

Some of the continuing problems have already been mentioned, such as those associated with the notices, the online catalog and the barcodes. Another that adversely affected the library-client relationship involved the Dynix user documentation. It was clear and well written, but incomplete. At the time of installation, only about 25% of the projected documentation was available. New chapters began to arrive with greater frequency during the following year, but some of the chapters missing were those that happened to be crucial to the areas of the recurring problems.

The most serious blow to the relationship between the library and Dynix occurred during the summer and fall of 1985. This concerned a major purchase of hardware to upgrade the computer. The library located a vendor other than Dynix who was able to supply the needed equipment at a substantially lower price, and placed the order. The Ultimate Corporation, however, was disposed to recognize Dynix as the sole vendor for the Smith Library, and therefore failed to honor the purchase placed through the secondary vendor. Instead, the order was sent unfilled to Dynix officials for their decision as to whether the order should be honored or not.

The library suspected that Dynix was delaying the order, and protested strenuously to the Dynix customer service representative handling the Smith Library account. The two vendors confronted each other over the issue and then gave conflicting reports to the library. In the wrangling that followed, the library's order for the needed equipment upgrade remained unprocessed for weeks. As a result, the automation staff was forced virtually to halt development of the system owing to a lack of disk space. Furthermore, the demand for greater access to the system by library users went unanswered, and response time reached unacceptably low levels. Eventually, the order for the equipment upgrade was processed, and Dynix honored the lower price quoted by the other vendor. A new disk, new ports and a faster processor were purchased. But the delay had upset the library's timetable for the planned growth of the system, and the library-vendor relationship sank to its lowest level. In consequence, the Dynix customer service representative assigned to the Smith Library was changed, and the general manager of Dynix came for a personal visit to the library partly to assess the deteriorating situation.

The visit occurred in November, 1985, at the library's request. For three days, the Dynix GM met with each department of the library separately to assess the performance of each of the Dynix modules. His response was very satisfactory. He corrected on the spot those software problems reported to him that could be corrected easily, and took notes and examples of those that needed further analysis by the Dynix programmers in Provo. He also itemized those problems that had already been corrected in a new software release that was forthcoming. Inasmuch as the notices program was working properly by that time, he simply needed to answer a few questions that had arisen as a result of the lack of documentation on that feature.

In addition, the Dynix GM wrote a program that greatly alleviated the problems that were occurring because of the incompatibility of the library's old barcodes. The number of barcode-related problems dropped virtually to zero after the visit. Furthermore, he clarified the Dynix position concerning the disagreement with the library over the earlier hardware purchase, and thereby restored much of the company's credibility in the matter.

The visit was profitable for both the library and Dynix. But it left the library staff wondering why the corrections that were so easily made had been so long in coming, and why the library had been receiving different messages and attitudes from other members of the Dynix staff than those held by their GM. The answer probably lay in the fact that Dynix was a new company experiencing astounding growth and success. In such a circumstance, it was necessary to expand the staff quickly and add persons with limited automation, library and business experience. Training the new staff members at Dynix must also have presented difficult problems. Indeed, during the library's first year with Dynix, no less than three different customer service representatives admitted over the telephone to the library's automation coordinator that they had only limited knowledge of the system. In one instance involving the lack of documentation on the notices feature, the customer service representative admitted ignorance in that area, and told the coordinator in all seriousness, "You'll just have to experiment."

It is also possible that this corporate newness was a factor in the way problems were communicated to the Dynix programmers from the customer service representatives. Instead of talking directly to the Dynix programmers, the library was able to speak only to the customer service department, which then relayed a description of the problems to the technical staff. Any breakdown in that communications path could contribute to a delay in obtaining a solution to a system problem, and would certainly account for the persistence of some of the problems encountered by the Smith Library.

It is important here to reiterate that the library actually had multiple vendors, even discounting the several vendors who supplied the peripheral equipment such as terminals, printers, light pens and so forth. The Ultimate Corporation in New Jersey and the local maintenance contractor were two separate firms, and neither knew anything about the Dynix application software. Dynix, of course, was an Ultimate dealer, and so the Dynix staff was familiar with the Ultimate system and tried to be helpful with respect to the operation of the computer. Yet each of the three main vendors was a separate entity with distinct responsibilities toward the library's system. There was not a clear explanation at the outset, written or otherwise, as to what those respective responsibilities were.

This situation is common with automated library systems. Only online experience during problem situations can help clarify the sometimes vaguely defined areas of vendor responsibilities. It is important, therefore, that a library with an in-house configuration have a substantial level of expertise available, especially during the period of time following the installation, when problems are likely to be numerous. The presence of the expertise may not bring about a rapid solution to system problems, but the solution will certainly come more quickly than if the expertise is lacking. At the Smith Library, the experience and know-how of the automation staff greatly accelerated the problem-solving process because the problems could be defined more quickly and communicated to the appropriate vendor. Nevertheless, the events described above show that even under the best of circumstances, problems can still be quite numerous and persistent.

LOCAL DEVELOPMENT

The in-house configuration at the Smith Library, then, made little difference in gaining control of the system as far as the vendor-supplied software was concerned. The library staff was still required to work through outside personnel, namely, the Dynix staff, for certain types of system maintenance just as it had been obligated to work through Computer Services when the system was hosted on the university's main computer. The in-house configuration did make a difference with respect to the level of computer expertise needed within the library. More expertise was necessary simply because Computer Services personnel were no longer performing any hardware or software operations for the library as had been the case previously. In addition, the expertise minimized serious problems with vendor software, such as creating the security shell around the public catalog and writing the program to encumber student files in the absence of a working notices feature.

When the in-house option was chosen, hardware expertise also became essential in two separate contexts. First, the automation staff had to grasp quickly the techniques of operating the computer under normal conditions. Different persons were assigned responsibilities for different functions, but everyone learned such things as how to configure the printers, operate the tape drive, clean the tape path, restart the computer after a power failure, search tapes online, create tapes, load data from tapes, perform file saves and file restores, set baud rates and other port settings, query the system and see to all the other functions that must be performed as part of normal day-to-day operation of the system.

Second, the expertise became a critical factor in unusual or emergency situations, such as a major hardware malfunction, broken file links and installing new releases from the Ultimate Corporation, all of which were occasionally troublesome. Such instances left the computer in a confusing state of disarray. Without the in-house expertise, the situation could have become badly muddled. The results would have required lengthy and tedious repairs either by the vendor via modem, or by the library staff as directed by a distant troubleshooter during a "talk through" over the telephone of very complex and confusing recovery procedures. In either case, recovery time would have been substantially lengthened without the presence of in-house expertise.

As discussed earlier, the library chose to exploit the local expertise and develop its system beyond the operation of the turnkey modules. The hybridization project was an enormous undertaking that would not have been possible without the in-house programmers. In order to benefit from the expertise available within the library, the new automation coordinator reorganized the automation department soon after the installation of the library's computer. New procedures were established to allow a much greater cross-flow of information and communication between the automation staff and the end users. Also, policy statements were written that governed departmental operations, and procedures were implemented to monitor the system's security and the performance of the equipment. An extensive tape library was established to ensure adequate system backup. Documentation for the locally developed modules was begun on two levels: (1) in-program documentation written specifically for programmers, that minimizes the adverse effects of student turnover, and (2) end-user documentation for library workers.

It was also necessary to ensure that all automation development take into account recognized practices of systems analysis and design, and that projects be guided by the principles of librarianship and addressed to the specific needs of the Smith Library. Taking the extra time to analyze the procedures being used and then design appropriate systems to automate them, instead of proceeding directly into the programming as had been done with the library's original system, provided a more solid foundation for the hybridization project. Planning proved to be just as important as programming, perhaps more so.

Placing a librarian in charge of the project instead of a technician or programmer was also a crucial ingredient in the success of the local development. Close supervision by professional librarians balanced and gave direction to the technical expertise of the non-librarian programmers. Working together, the automation staff was able to gain momentum and press forward quickly and effectively. The files were carefully designed to library needs and standards, and properly constructed to gain the most benefit from the computer's operating system. The programming was clean, well-structured, and efficient according to the most current methods of applications programming, and was also effective with respect to the library's needs.

At present, the locally developed modules are nearing completion, and are in fact in operation at the levels to which they have been completed. Because of their design, different segments of each module can be used as they are developed. The modules include media scheduling, acquisitions, reserves, inventory control for audio-visual hardware and an online tutorial. In addition, a number of utility functions are being developed to aid in system security, COM production, remote access via modem and in other areas. All the files were designed and linked in such a way that the major modules are fully integrated and share common data files with the Dynix modules.

After several years, then, and some significant changes in direction, the Smith Library has a system that will automate all of its major functions and services. To arrive at this point, the library has gone through a number of phases that approximate the course of library automation in general. From a locally developed, remote host system that began as a circulation control system, the library moved to a larger, rather expensive and inefficient system, and then to a turnkey package with an in-house configuration. Then the library took what looked like a step backward toward local development again. The result is the present hybrid system. Hybridization may well be the next logical step for library automation in general.

The Smith Library found much of its expertise in the relatively inexpensive student programmers from the university's computer science department. By coordinating the technological resources at hand, the library was able to draw from the structure of its organizational environment the precise elements needed to achieve its automation goals. Other libraries that venture into local development will be in somewhat different situations, but each will be part of a political and technological environment in which may be found unsuspected resources that can be called upon when automation is wanted. In any such project, however, it is imperative that the endeavor be directed by a librarian and that communicational paths and organizational flexibility be established.

RECENT HISTORY

The relationship between the library and Dynix improved greatly as a result of the visit by the Dynix general manager. Not only did he solve some of the software problems immediately, but he also saw to it that other problems received attention when he returned home. Furthermore, he resolved many of the communications difficulties the library had experienced in the past when dealing with Dynix. Procedures at Dynix headquarters were streamlined and organizational improvements were made. The cumulative effect of these actions were a faster response to the library's requests for assistance and a restoration of mutual trust and commitment between the library and Dynix. Problems are now resolved much more quickly, and the library's position with respect to obtaining hardware at the lowest possible price is understood. An aggressive program to complete the system documentation is also underway at Dynix.

A massive new software release was installed on the library system in April, 1985, and this has solved most of the problems that were lingering from the time of the original Dynix installation. It is unfortunate that this new release was so long in coming. Its delay not only obliged the Smith Library and other Dynix clients to put up with certain system problems for an unreasonable period of time, but created a number of hardships during installation of the massive new software package. For example, the Smith Library system was down for two days, and a significant amount of retraining was necessary to acquaint the library staff with the many software changes.

The awkward size and complexity of the new release prompted Dynix to establish a policy of issuing smaller, less involved releases more frequently instead of stockpiling corrections and enhancements for a single, large upgrade to be issued only every year or so. This approach is more in accordance with industry standards and promises to be much more satisfactory to Dynix as well as its clients.

The new release included numerous enhancements to an already highly sophisticated system. It is not appropriate here to enumerate the Dynix features except to say that system sophistication was a major interest of the Smith Library when a turnkey package was selected. With the new release, the features that originally made Dynix attractive have increased. So, too, has increased accordingly the importance of staff expertise to handle such a powerful system effectively.

T. D. Webb
D. Errol Miller

REFERENCES

1. Joseph R. Matthews, "Unrelenting Change: The 1984 Automated Library System Marketplace," *Library Journal* 110(6) (April 1, 1985):33.

2. Matthews, p. 32.

3. Richard De Gennaro, "Integrated Online Library Systems: Perspectives, Perceptions, & Practicalities," *Library Journal* 110(2) (February 1, 1985):38–39.

Chapter 6

Analysis of the Case Studies: Host Location and "Decision Layers"

The Phoenix Public Library and the Joseph F. Smith Library are at different extremes of several scales. Most noticeable are the scales of size and type. With nearly 1.5 million volumes, Phoenix Public is a major urban research library, while the Smith Library is a medium-sized collection one-tenth that size. As two different library types—public and academic—they also have differing characteristics with respect to user types, services provided, collection development, and staffing, to name only a few. The case studies demonstrate other crucial differences including degrees of resident computer expertise, goals of the respective automation projects and location configurations of host computers.

These and other differences greatly affected the libraries' selection of their respective automated systems and the organizational designs necessary to administer them. The degree of institutional choice allowed each library by its parent organization was a further determining factor. Phoenix Public, like many libraries, was required to locate its computer at a remote site, while the Smith Library was free to locate in house.

Since the thought behind this book is dependent upon a library's ability to make a choice about host location, it may seem that if a library has no choice but to locate its computer where its parent organization decrees, an evaluative comparison of remote host and in-house configurations is moot. Such a comparison does, however, acquaint the library with what it has to look forward to in either case. A library may, of course, have more choice in the matter than may be initially apparent. Deployment of computer resources throughout large organizations is becoming much more common. With virtually all organizational levels and functions undergoing computerization, centralized computer facilities are finding it difficult to keep the ubiquitous machines under their thumbs. A library administrator who ordinarily would be required to submit to a remote host policy may be able to secure an exception by (1) demonstrating the growing trend toward in-house configurations among libraries and the prevalence of systems designed for in-house operation, (2) exerting persuasion and political leverage to obtain special favors for the library, and (3) assuring his or her superiors and the computer experts that the system will be operated and maintained in a competent manner with regular consultation from the computer center.

Computer competence is the fundamental concern, and the library manager should ensure its availability before ever considering an in-house configuration. The case studies demonstrate that beneath the differing library-specific conditions of automation projects lies a uniform series of potential risks that require a high degree of expertise to master.

Admittedly, there are widely varying levels of quality available in turnkey software, hardware, training and maintenance. Some of the problems encountered by the libraries in the case studies can undoubtedly be laid to the fact that their vendors were new. Other vendors would have presented these same libraries with a somewhat different array of conditions and problems. Yet one has only to browse through a few current issues of any professional journal to see that the problems encountered by Phoenix Public and by the Smith Library are typical of problems encountered in libraries across the nation. For example, Hegarty reports,

> In the state of Washington, there are four libraries that purchased automated systems at approximately the same time three years ago. The selected systems are still not ready for final acceptance testing. This has nothing to do

with the relative merits of any particular automated system, but rather that the acquisition of any automated system brings with it a burden of responsibility and maintenance that is not trivial.

He then lays to rest the myth that "You don't have to be computer literate to make effective use of your automated system":

> It is imperative that library staffs become computer literate. It is not enough that they have a basic understanding of what computers can do. They must also develop a proficiency and a confidence level. These can only come through hands-on experience. . . . As our staffs grow in computer literacy, we then become capable and equipped to bring pressure on the vendors to enhance their systems to state of the art levels. This can only come through constant training, retraining, education, and reeducation of our staffs.[1]

The case studies indicate, first, that a high level of computer expertise is necessary to operate turnkey systems efficiently, with even higher levels required for any expansion of system capabilities beyond the turnkey features. And second, the effective use of computers by librarians at present depends not so much on the location of the host as on shrewd administrative and organizational control of the technological and political circumstances surrounding the library.

Neither the administrative nor the technical worries of running an automated turnkey system disappear when the computer is moved in house. On the contrary, they become the direct responsibility of the library. Despite the obvious organizational and communications headaches involved, MIS relieved Phoenix Public of the complicated tasks of maintaining the system, which included the normal day-to-day operation of the machine and handling the emergencies and catastrophes such as those that occurred in the summer of 1983. These would have been even more longlasting and debilitating had the computer been located in house and maintained by non-experts at the library.

If Phoenix Public had located its computer in house, the library would have had to expand its staff significantly with programmers, systems analysts and operators, just as the Smith Library did. Further, it would have been necessary to obtain an expert to supervise the project. With all this specialized expertise on board, a reorganization of the library's operational structure and the establishment of a separate organizational unit to maintain the system could not have been avoided, again just as the Smith Library discovered.

All these operations were performed for Phoenix Public by MIS, which had the required organization and personnel already in place. Indeed there were instances of gratifying success at Phoenix Public when the technical expertise of MIS and the administrative adroitness of the library were properly combined. At the Smith Library, the required computer operations were performed by the library's own automation department. The case study records what some of those operations were and, in doing so, provides added appreciation for MIS at Phoenix and for other similar departments, where the computer expertise of a government structure resides. At the Smith Library, moreover, it was necessary to give careful direction to the work of a group of experts, and thereby control their efforts for the benefit of the library.

The direction of the automation project at the Smith Library was quite different from that of the Phoenix Public project, and the expertise required was a good deal more extensive. This only increased the amount of careful supervision needed. Increasing expertise requires a corresponding increase in managerial skill. According to Howden,

> Library managers are not foolish enough to expect librarians to be universalists and certainly they would not expect programmers to do every task equally well. In making assumptions about programmers there is always the possibility, however, that the programming ability will be equated to an ability to design subsystems. In any library environment programmers can be expected to perform as well as we can match them to the tasks to be performed.[2]

He also says,

> Regardless of the library knowledge a programmer may have, I place far more confidence in someone to whom I can communicate a problem and who will work single-mindedly to solve it. A truly hungry programmer I will free from administrative constraint, provide with a private terminal, and leave alone. Once the problem has been rigorously outlined, I know the product will arrive early and have capabilities that will be exceptional.[3]

Paradoxically, in order to build this kind of flexibility into an automation project, it is necessary to increase the complexity of the organization by creating a separate unit the only responsibility of which is development and maintenance of the system. The Smith Library is a good example of the type of reorganization as well as of the increase in organizational complexity and formalization that accompanies an aggressive automation project. Although automation affects all areas of the library, the unit charged with the project must be organizationally free from the other units in the library so that it can interact with them more openly, establishing a flexible staff relationship with them unfettered by line responsibility to them.

At the same time, Howden warns against the use of student programmers in whom he finds "gaps in understanding of job control language, customer needs, and timesharing techniques." He finds them generally "naive in business practices, organizational skills, and customer applications." He argues that libraries intending any local development will probably need programmers with an unusually high level of expertise. This is because (1) library automation involves some peculiar operations not found in scientific or industrial programming, in which most computer science students receive their training, and (2) libraries usually do not have the resources or senior personnel to provide training to upgrade student programmers.

Library automation, he says,

> requires the use of file structures that are often quite sophisticated such as IBM's VSAM technique. File sizes are usu-

ally large. Data is record-oriented with variable length fields and variable length records. Much processing deals with character rather than numeric data. Some techniques are relatively unique to the field, such as inverting files, sorting in unusual patterns, and building subject search retrieval systems.[4]

But the Smith Library case study demonstrates conclusively that with the proper training and supervision, student programmers can master the sophisticated techniques of library automation. Thus they can fill the role of computer technicians to maintain a turnkey system as well as perform the programming and analysis necessary for local development.

It appears that local development to one degree or another will be required if libraries are to obtain the types of systems librarians seem to want. Librarians are crowding vendors for enhancements and features that are not likely ever to be part of a turnkey system. Epstein compiled a list of over twenty functions that librarians reportedly want to see in their integrated systems.[5] In addition to the standard bibliographic functions such as acquisition, cataloging, public access catalog, circulation, authority control, and so forth, the list includes such unlikely things as personnel records, staff schedules, payroll and check production, teletext, and cable television links.

Most of these functions will never be incorporated into a turnkey system because the demand is not sufficient or uniform enough to be worth the vendor's development time. To develop and market them would price a system out of the market. But obviously, some libraries do want to automate these functions. De Gennaro advises that instead of pressing vendors to build these ancillary functions into a turnkey system, "Librarians would be better advised to buy the standard commercial programs for a few hundred dollars and run them on microcomputers."[6] Going a step further, it will soon even be possible to network several micros so that different work stations in the library will have simultaneous access to, say, the information and referral files, budgetary files, outside databases, and so forth.

Although some libraries are in fact moving in this direction, there are disadvantages to automating library-specific files on a micro system. First, while handy for use by trained staff, a micro does not lend itself well to public access. Damage to the machine and to the data can easily occur if made available to the public. Yet some features, such as ready reference files, would be of great value for direct public access. Second, the micro system would not be integrated with the main system. It conceivably could be necessary to duplicate some of the data in both systems. Duplicate effort to search them both in certain instances would therefore be necessary.

Another alternative method of automating these specialties is through local development of files and software on the main system in a hybridization project like that at the Smith Library. But as the case study shows, this would require that an even greater level of computer expertise be at the disposal of the library. It seems, however, to be the only way to get a tailor-made system.

The next phase of library automation may well involve systems "specially designed" by the vendors to allow local hybridization, as if that potential had not always been present.

Despite their inability to cater to every need of every library, vendors can be expected to provide increasingly sophisticated systems. There are two reasons for this inevitable technological escalation. The first is the natural development of the computer industry. New hardware and software and new methods of building and developing systems will certainly find their way into library automation. If older vendors are unable to provide the new items, then vendors will appear who can. In either case, it means that libraries in the future will have an ever-increasing need for access to computer expertise, regardless of host location.

Secondly, as mentioned earlier, librarians themselves persistently request more sophistication, or at least so go the claims of the vendors. In actuality, many librarians are probably satisfied with current levels of sophistication. Many are overwhelmed by them. Yet vendors who market more powerful systems with more advanced searching capabilities and larger numbers of on-line functions always seem to find clients, thus urging other vendors to even greater feats. There seems to be no end to the spiral.

Libraries that purchase the newer systems become what De Gennaro calls "pioneers":

> Anybody who buys such a system should be prepared to experience all the pains and pleasures of pioneers who, as we all learn in school, suffer terrible hardships before they get to the promised land — and many never get there. [7]

Phoenix Public and the Smith Library both qualify as pioneers. They suffered because of the inexperience of their vendors, but made contributions to the gradual improvement of their vendor's operations and products. And the libraries in turn benefited from enhancements to the system. The library community at large is sure to benefit from this feedback cycle to the vendors and system designers because the demand for enhancements will have a cumulative effect in the library automation industry. An example is the astonishing increase in power and sophistication of the Dynix system compared to the early ULISYS system, both of which were considered state of the art products at the time they were purchased by the libraries in the case studies.

The cumulative improvement of library systems, however, will occur only if libraries are firm and united in dealing with the vendors. As Matthews says, "The marketplace dynamics say that vendors will only change when it's clear that changing will get them more business. It's up to librarians to seize control of the process."[8] He adds, in an observation about who currently has control,

> The old saying, "once they gotcha, they really gotcha" is applicable here . . . Vendors need to be more concerned about establishing and maintaining a "good" reputation in the marketplace than in short-term sales. [9]

That "old saying" was in fact particularly appropriate to the Smith Library situation where Dynix took a year and longer to respond to some of the problems in the circulation and public catalog modules. After repeated telephone and written appeals that the bugs be removed, the library staff finally gave up asking and resigned themselves to living with them.

The library preparing to automate has many defenses against unfair treatment by vendors. These include such things as (1) issuing a well-planned and thorough request for proposals (RFP), which establishes early in the automation process those requirements a vendor will be expected to meet; (2) designing a contract that protects the library's interests more closely than the "standard" contract the vendor is likely to submit; (3) establishing an equitable but thorough testing/payment schedule before installation; (4) talking to as many of the vendor's clients as possible, by telephone or in person, as part of the selection process.

The last mentioned recourse is particularly important, but should be handled verbally, not by written correspondence. People are likely to speak more candidly than they write, and a verbal exchange allows the freedom of discussion necessary for probing interviews and clarification of meaning. Unfortunately, neither Phoenix Public nor the Smith Library were able to take full advantage of this strategy because neither Universal nor Dynix had a sufficient number of clients at the time to provide a balanced survey.

Solving or preventing interagency crises of the type experienced by Phoenix Public requires strategies that are essentially political or interpersonal in nature. It requires finding a way to span the interagency barriers and navigate around blockages in communication channels or in the organizational structure. As the case study from Phoenix Public indicates, these maneuvers are possible and can yield great success. But interagency cooperation requires a blend of patience and judicious forcefulness that may be almost as scarce as computer expertise.

This is not the place to discuss all the procedures by which a library can seize control of its automation project from either a vendor or a neighbor agency. Many of the procedures that deal with library-vendor relations are adequately and admirably addressed elsewhere in the literature, which should be studied thoroughly prior to any automation project.

The case studies demonstrate that library automation is, above all, a management challenge. Computer expertise is essential and fundamental to any automation project. But the Smith Library learned that even an abundance of in-house expertise cannot always resolve problems or prevent them. Phoenix Public learned that expertise is of no value if it is administratively, organizationally or dispositionally inaccessible.

Expertise is vital to an automation project. It will become even more so in the future. But it must be managed and directed. Avenues for its use must be opened and broadened on the local level between the units of an automating library, and in the larger environment, which may include vendors, consultants, and system operators at a remote site.

Underlying the matter of host location, then, which many libraries only casually consider, and which many others consider from an erroneous perspective, are layers of other considerations that are never broached. These decision layers involve a search for the expertise necessary to handle a potential system, a search that, like any other sort of resource acquisition, is a managerial function: where is it? who has it? how much will it cost? how can the library acquire it in sufficient quantity and quality to operate the automated system at maximum levels of efficiency and effectiveness? once located, how can the expertise be controlled?

If a library manager chooses to locate the host computer in house on the basis that such a configuration will allow better control of the system, and yet is unable to find a librarian with the expertise to operate it, the advantage of having the computer close at hand is lost. While the staff may be able to operate the vendor's application software reasonably well, it will not be able to control the system itself with the optimum degree of efficiency.

The library manager, then, has two possible courses of action. First, the computer may be located at a remote site to be operated by computer experts with no library background. If this course is chosen, that manager's task becomes one of controlling and directing the expertise even though it is part of a separate organizational unit. The case study from Phoenix Public Library illustrates the pitfalls and the potential successes that are possible by following this course of action.

Otherwise, the library manager may choose to locate the computer in house. In this case, the problem once again becomes one of controlling expertise that, though organizationally close, remains professionally distant. The case study from the Smith Library illustrates the success that may come from this configuration, but it also dispels many of the illusions that library managers may entertain concerning an in-house computer and its operation.

REFERENCES

1. Kevin Hegarty, "Myths of Library Automation," *Library Journal* 110(16) (October 1, 1985):48–49.

2. Norman Howden, "Programmers and Inverted Files," in *Conference on Integrated Online Library Systems. September 26-27. 1983: Proceedings*, ed. by David Genaway, rev. ed. (Canfield, OH: Genaway & Associates, 1984), p. 278.

3. Howden, p. 275.

4. Howden, pp. 275–276.

5. Susan Baerg Epstein, "Integrated Systems: Dream vs. Reality," *Library Journal* 109(12) (July 1984):1302–1303.

6. Richard De Gennaro, "Integrated Online Library Systems: Perspectives, Perceptions, & Practicalities," *Library Journal* 110(2) (February 1, 1985):38.

7. De Gennaro, p. 38.

8. Joseph R. Matthews, "Turnkey Systems: High Risk for Libraries?" *Library Journal* 110(4) (September 1, 1985):135.

9. Matthews, p. 135.

Chapter 7

The Professional Issues: Conclusions and Recommendations

In simple terms, the conclusion of this book with respect to host computer location is that if a library chooses the in-house option, that library needs in-house computer expertise. If the expertise cannot be obtained, the recommendation is that a remote host option is probably a better bet. An in-house configuration does not by itself guarantee effective control of the system, but it greatly increases the need for a qualified staff to operate the system effectively.

To the degree that computer expertise is in short supply across the profession, steps must be taken to alleviate the situation. Otherwise, librarians may lose control of their profession. This condition gives rise to the professional issues of library automation. As discussed in the preface, this book hopes to illuminate several professional issues of library automation that tend to converge in a meaningful way in the matter of host computer location. Since significance of the issues is determined by the circumstances in which they come to bear, circumstances different from those associated with minicomputer operation would afford somewhat different meanings to the issues. For instance, the increasing automation of reference services may well give additional significance to professional issues other than those specifically addressed here. Nevertheless, it is likely that the broad impact of automation on librarianship as a profession has created a number of conditions of which the present discussion is generally indicative.

The preceding chapters indicate the circumstances that give particular urgency to the several issues mentioned. It is appropriate now, therefore, to re-examine each of those issues separately in light of the findings of this study.

LIBRARIANS AND COMPUTER LITERACY

A recent survey of 1,400 members of the American Library Association indicates that a majority of practicing librarians feel overwhelmingly a need for computer and technological training:

> Computer and technology training and management development training (59% and 50% respectively) were most frequently mentioned as areas in which professional development is needed. The need for interpersonal skill training and for subject training (science, business, language, etc.) was noted by 20% of the total sample. Fourteen percent cited career planning as a professional need.[1]

The degree to which librarians feel the need for computer training, ranking it significantly above even management training, indicates clearly the extent of the deficiency in technical know-how. The survey results tend to substantiate the claims of this study that computer expertise among librarians is marginal and may be inadequate to manage the minicomputer-based systems now common in the library marketplace.

A library can certainly locate its computer in house and rely on the vendor to train the staff to perform the standard operating procedures. Libraries are doing this every day. But the results too often are minimal efficiency and effectiveness. Library staff members generally lack sufficient expertise, and the vendors are unable to provide it. According to Matthews,

> Vendors usually offer one-time training sessions instead of focusing on the ongoing training needs of the library.
> While it is true that a vendor's primary responsibility in this area has traditionally been to train a library's trainers (who in turn train other staff members), this approach is terribly short-sighted. . . . Too many vendors and librarians assume that once training is presented, all library staff will be conversant with the system. This is simply not true.

Many libraries are only enjoying 25 to 50 percent of the potential benefits of an automated system because their current staff members have received little or no training that has real value.[2]

Sloan warns that

Poorly trained people use systems poorly, if at all. At best, poorly trained personnel represent wasted fiscal resources, using a system ineffectively, or not using the system at all. At worst, poorly trained personnel make mistakes that interfere with the operations of the organization.[3]

Librarians may or may not someday have the expertise themselves in sufficient quantity and quality to get more than the minimum from their systems. But this is off in the future, and will require massive amounts of re-education and a careful redirection of library education and professional recruitment. The question is, Can libraries in the interim afford to be satisfied with minimal returns from their expensive, sophisticated systems? The answer is that they cannot. Buying a turnkey system just to let it run in low gear is not adequate for the feats of information delivery libraries will be required to perform in the future.

Ideally, of course, an automated library system should be geared to the literacy level of the staff that will be using and operating it. Any library that is planning to automate should carefully evaluate the degree of computer literacy resident among its staff. Clearly, however, a greater concern and the final aim is to get the best return from the system once it is installed. It may be that in some instances, this will require going outside the profession in one manner or another to obtain the necessary expertise.

As the case study from the Phoenix Public Library shows, going outside the profession, or even outside the library, can represent an efficiency trade-off. That is, in order to obtain greater automation expertise, the library faces the risks of communicating through inter-organizational channels that can be convoluted and highly inefficient. In the Smith Library, likewise, although the system and its operation were located in house, the automation project resulted in increased formalization and complexity in the library's organization.

At issue, then, with respect to computer literacy among librarians, are the methods of maintaining sure and adequate control of library systems, despite inadequacies with respect to computer expertise. Ensuring proper control will no doubt require changes in certain aspects of a library's procedures, operations, and organization, and each library will present a different set of circumstances. However this may be, control of the expertise, if not outright possession of it, is the essence of the professional challenge inherent in library automation.

LIBRARIANS AND SYSTEMS DESIGN

McGee states,

> Even today, when libraries with unusual requirements attempt to obtain suitable systems from the turnkey market, there is sometimes a reluctance or inability on the part of vendors to appreciate the differences between their products and the libraries' stated needs.

He says further,

> It is no surprise that the perfect library system has not been developed. Besides failing to understand libraries' requirements, vendors often develop capabilities that are convenient to provide, rather than truly responsive to genuine library requirements. Such shortcuts have led to costly mistakes for both vendors and customers.[4]

Obviously, control of an automated project begins with its design and development. And just as obvious is the fact that the nature of turnkey systems precludes the library, as a potential customer, from having any say in the system's design phase. In actuality, a library can "design" very little, if any, of the features of its prospective turnkey system, insistent vendor claims to the contrary notwithstanding.

The library automation marketplace, however, is subject to drastic and rapid change because it is linked to computer technology, which is still a relatively new and rudimentary field. The greatest promise for library automation, therefore, lies not in learning how to handle present systems and machines, which are generally awkward and unfriendly, but in participating fully in the design of future systems.

Library literature just now (and presumably library thinking) is a glut of evaluations and advice about how to get along with systems that are currently on the market. We monitor systems to find the best and the cheapest, and spread the word about these in print and through the grapevine; we keep track of the aggressive vendors to find out who has the biggest share of the market, thinking that this might reflect system quality; we bury ourselves under encyclopedic monographs that tell us how to automate our libraries. Of course, this type of information is vital, if somewhat misleading and overabundant, because libraries need to be especially thoughtful and selective during this initial period of library automation. The fact remains that present systems must suffice for now.

The major issue, however, is provision for the future. Myopia with regard to technology tends to distort our perception of constraints that will be placed on us in the future if we fail to take part in the development of new technologies. No library should expect to stay with an automated system "for life." If the fields of scientific research and business are any indication, libraries should expect to change systems with some frequency either through major upgrades or outright replacement. The case studies included here demonstrate this inevitability. They demonstrate likewise that the new systems and technologies of the future will be much more useful to libraries if librarians are prepared to contribute to their design.

I have found that a thorough understanding of such things as system architecture, file structure and linking techniques, and a good grounding in systems analysis and design, even more than actual programming ability, have been most helpful in efficient and effective use of automated systems, whether these are of the in-house or remote host configuration. And I believe these to be the most valuable skills the profession can cultivate with respect to the future of library automation.

As discussed earlier, computers are still rudimentary and unfriendly to the average person. Nevertheless, the flexibility, capacity and friendliness of computers have vastly increased in a relatively brief period of time, and further improvements will no doubt be forthcoming soon. The changing nature of computer technology suggests that when libraries as a class have mastered the skills of systems analysis and design, the time will be favorable for reconsidering the local development of automated library systems. This option would normally be restricted to large, well-funded libraries, and perhaps to those libraries that have access to good, relatively inexpensive student programmers to do the actual programming. As the Smith Library case study makes clear, excellent student programmers are available. If these students are carefully supervised according to the principles of librarianship and the practices and needs of a particular library, and if their work is combined with a knowledgeable application of systems design and analysis techniques, a locally developed automated system becomes a more feasible possibility now than it was at any time in the past.

The proliferation of turnkey systems and the prohibitive costs and time delays associated in the past with locally developed systems tend to obscure the growing feasibility of the local system. But as the technical difficulties are resolved through development of better machines, libraries are bound to start developing their own systems and sharing these within the profession. In cases of this sort, it is hoped that librarians will be the designers of the systems, but not have to muddle in the programming, which can be very tedious, time-consuming and expensive and which actually is not related to system architecture and design. Again, however, this approach requires precise communications skills and organizational flexibility within a library, and considerable systems experience among the librarians involved.

AUTOMATION AND LIBRARY PHILOSOPHY

There is a question as to whether or not a theory of librarianship and an accompanying philosophy, separate and distinct from that of other disciplines, actually exist. Freeman fears that librarianship, because it borrows theories so freely from other disciplines, may lack its own set of theories, and in fact "reflects the failure of librarians to rationalize phenomena by the application of principles and strategies by their peers."[5]

On the other hand, Buckland argues that

> each of the various parts of the theory of library service is probably not unique to library service or, at most, is unique in detail of application only. Further . . . this very lack of uniqueness would seem beneficial from a practical point of view because it permits librarians to collaborate with others on shared theoretical problems. At the same time, there would appear to be no reason not to regard the totality or combination of the theoretical aspects of library service as being unique to librarianship. On reflection, this could seem to be an ideal outcome: Librarians can take pride in having something unique and take advantage of the fact that they share so much with others.[6]

Of the philosophy of librarianship, Buckland says that "philosophy—in the sense of systems of motivating beliefs, concepts, and principles—must necessarily pervade the provision of library services."[7]

This is not the place to do any more than mention the controversy and say, in partial agreement with Buckland, that to whatever extent a philosophy and theory of librarianship exist, they are manifest in library services, and "that library service, the use of technology and skills, is and must necessarily be deeply value-laden, in the sense that uses are involved which relate positively or negatively to social values."[8]

To date, the effect of automation on the services, theories, and philosophy of librarianship has not been so much to alter them, but to re-endorse them and the social values they imply. Computers have been the proverbial "shot in the arm" for the profession. Automating library processes has produced a focus of attention on library services in exquisite and often excruciating detail. Yet this heightened clarity of self-examination has yielded few new ideas to speak of or, in Freeman's terms, any new, systematized, long-awaited theories "that explain, predict, and describe the structure of information and the dynamics of library management."[9]

Far from lamentable, this instead demonstrates that there are in fact guiding principles of librarianship, if not an outrightly articulated philosophy, which have so far remained more or less constant amid all sorts of bewildering transformations. These guiding principles include such things as the conviction that a free society depends upon unrestricted access to knowledge and information and a commitment to unfettered public service. The fierceness with which librarians commonly defend these pursuits certainly seem to bear the force of philosophy, even if they are nowhere systematically structured into philosophical formalism.

Any number of persons or organizations, of course, defend and in some way give actuality to these same principles. According to Buckland, librarianship achieves its uniqueness in the way it applies and administers them, that is, in the way the vehicle for disseminating knowledge and information is organized and managed. And with respect to the way in which services are delivered, automation will have a significant effect on libraries.

As early as 1965, in an early work on library automation, Wasserman reported, "Mechanization advances the acceleration of the formalization of the organization, resulting in a furthering of the rationalization of work brought about through the substitution of rules for judgment."[10] This is because machine procedures are exact and inflexible. Once automated, a library must submit any number of procedures that once were informal to the formal procedures required by machine accommodations. But Hall makes an interesting observation:

> Studies of professionals in organizations have come up with quite consistent findings: The organizations involved are less formalized than organizations without professionals. As the level of professionalization of the employees increases, the level of formalization decreases . . . The presence of professionals appears to cause a diminished need for formalized rules and procedures. Since professionals have internalized norms and standards, the imposition of organizational requirements is not only unnecessary, it is likely to lead to professional-organizational conflict.[11]

On the subject of organizational complexity, Hall likewise states that despite the intense division of labor common within an organization of professionals, the complexity is "a consequence of the differentiation of activities among the professionals," and is not organizationally imposed. Instead, the model followed in professional organizations, he says, is "collegial, which is characterized by formal distinctions or norms and by little in the way of formalized vertical or horizontal differentiation."[12]

Both case studies presented here indicate that automation greatly increases a library's organizational complexity. Horizontal complexity increases as computer specialists — librarians or otherwise — become part of the organization and ply their skills, skills that remain largely unfamiliar to the rest of the organization. Vertical complexity increases because the automation coordinator becomes another source of supervision with respect to the way library operations must be performed in order to assure their amenability to the computer.

Thus, although the philosophical principles of librarianship have remained intact so far during the process of library automation, we may well be on the verge of a major professional reconfiguration brought about by the increased organizational formalization and complexity that automation entails. If the uniqueness of library theory and philosophy lies primarily, as Buckland says, in the details of its application and administration of theories held widely in other disciplines, as automation changes those details, so will change the profession and its underlying philosophy.

ATTITUDES OF LIBRARIANS

A professional philosophy springs, of course, from the persons who practice the profession. It is their attitudes and backgrounds that contribute to the formulation of the professional philosophy. Rothstein's accumulated evidence that librarians "form a very distinctive group with respect to personality,"[13] along with the general and widespread commitment to the philosophical principles mentioned above, suggest that the attitudes of librarians are equally distinctive. This study has described two attitudinal aspects of library professionals that can be called personal-professional attitudes and computer attitudes that must be correlated. Changes in one set of attitudes portend changes in the other.

Summarizing a number of attitudinal studies of librarians, Rothstein states that as individuals, librarians tend to be critical, easily upset, sober, suspicious, apprehensive, anxious, undisciplined and tense. They are not outgoing, calm, venturesome, or experimenting. He surmises that librarians are "querulous loners," with a large dose of self-doubt, and concludes that librarians became librarians largely "because we had hoped to escape stress."[14]

Rothstein's arguments invite disputation. Nevertheless, to the degree that librarians do have a high anxiety rating, interaction with computers can be difficult for them. According to Baumgarte,

> Fear of computers reflects a generalized fear of current technology and is most prevalent in individuals who are highly anxious to begin with and forced to deal with computers especially in subordinate roles. The highly rational functioning of a computer can be very dehumanizing to a person whose job-tasks or class schedule is dictated by a computer. Such emotional reactions can result in attempts at subversion, absenteeism, and feelings of alienation.[15]

And lest we think he is referring to someone other than librarians, Baumgarte says further that "a third of all people in the information management occupations react with wariness to the arrival of a computer in the work setting and perhaps for a third of these the anxiety can be debilitating."[16]

Some people would argue that librarians are not really professionals, but are in fact "semi-professionals." According to Etzioni, semi-professionalism is characterized by (1) a relatively short professional training period, (2) an absence of life-and-death issues connected with the duties performed, and (3) a concern more with the communication of knowledge than with its creation or application.[17] Such a description is not entirely unrepresentative of librarianship. In the first place, the professional training is brief compared with that of physicians or attorneys.

Furthermore, Rothstein speculates that librarians are drawn to their profession because of its image as an "undemanding profession."[18] And Freeman observes that "librarians are great compilers of data, but not theory builders."[19]

If this is so, it may be that librarians are less affected by the professional-organizational conflict that, according to Hall, results from the increased formalization of libraries brought about by automation. Thus, they may have an adaptive advantage in one respect to automation.

The issue then becomes, How are the two sets of attitudes related, and how do they affect each other? If we change the attitude of librarians toward computers, which in light of this study seems a highly desirable goal, how will the attitudes that draw librarians to the profession in the first place be affected? And perhaps more importantly, how can the professional philosophy of librarianship, which in some respects is a summation of the mutual attitudes of individual librarians, be preserved while improving the computer attitudes among all librarians?

PROFESSIONAL RECRUITMENT AND EDUCATION

To preserve the unity of the profession while maintaining control of library automation in general, library science should balance automation and expertise. This means that librarians must do a lot of "catching up" with respect to their computer skills because it is not likely that the trend toward greater power and sophistication in library systems is going to stop. What is being proposed here, then, is a long-range method to raise the basal level of computer expertise in the profession.

There are two ways to rectify the imbalance between the modal level of computer expertise in the profession and the expertise needed to run an automated system effectively. The first is to attract recruits to the profession who already possess the desired computer attitudes and training. Given the commonly held image of librarianship, which Rothstein describes[20] and which so far has succeeded in attracting to the profession persons with little mechanical proclivity, this could be difficult. But the inflow of computer experts is certainly increasing, and as more of them become intrigued by the peculiar challenges involved in library automation, the number will likely increase even more.

The second way is to make certain that adequate automation training is provided in the graduate library schools for those who, on their arrival, lack computer expertise. Given the pervasiveness of automation in standard library operations, computer training should be required of all MLS candidates. Perhaps minimum standards of computer literacy and even computer science should be established. This may require some library schools to establish joint programs with the computer science departments of their universities or attach computer scientists as adjunct members to the faculty of the library school.

Both methods — increasing the inflow of experts and upgrading library education — are necessary to give librarians control over their systems. Although the methods for achieving these objectives are unclear at present, the result will be the "new breed" of librarian that Wasserman anticipated over twenty years ago.

The issue here is, Will the new breed ever appear? If so, it will only be as a result of the increased professionalization of librarianship. That is, the time required to complete a curriculum in library science should be increased to include training in the technology that has taken over the libraries where the library school graduates will be practicing their profession. The long-range answer to the dilemmas produced by library automation is not to relegate the responsibility for automation to the technician or the non-librarian computer expert, although these individuals are extremely important at present and will remain so for some time to come. To continue to rely on this source of expertise permanently, however, would constitute a gross error as it removes control of automation from those whose philosophical principles direct the library and its services in the first place.

Increased professionalization in the area of automation, however, carries dangers of its own. If Hall is correct, it can increase the possibility of professional-organizational conflict in libraries. Furthermore, it can change the content and appearance of the profession in ways that are as yet unclear.

Express recruitment practices, then, along with increased professionalization, may themselves threaten the philosophical principles that have made libraries the grand, if sometimes lumbering, institutions they are. Along with the increased professionalization, therefore, should be a commensurate reinforcement in the minds of librarians and potential librarians of the philosophical principles on which the profession is grounded.

Under these circumstances, increased professionalization appears to be the best method for proceeding in terms of library education. But it also has enormous implications for professional library associations. They must map the best route for practicing librarians to become more proficient in machine matters and provide some portion of the means to that end. This involves much more than furnishing conference meetings on microcomputing and arranging field trips to demonstration sites. It involves providing leadership for the refitting of the profession, establishing various types of performance measures and standards, and seeing to the upgrading of computer training in library schools.

Also required are carefully designed and administered in-service programs. According to the survey of Bernstein and Leach,

> For professional development, librarians would most likely participate in workshops and seminars (71% of the total sample). Professional association activities and reading professional journals ranked second (38% and 35% respectively). Regular university courses were noted by 24% of the sample; home study programs and computerized, programmed instruction were selected by 10% and 8% respectively.[21]

The popularity of workshops for increasing computer skills among librarians is widespread. Rappaport gives an example:

> The New York Public Library currently has over 40 micros scattered throughout the system. At least 300 people have received training and many more have had exposure. The numbers of computer literate people continue to grow because an environment has been created where more and more staff request computers and wish to learn. There has been a complete turnaround in attitude in the last two years.[22]

These seem like laudable and encouraging results. But based on the findings of this study, our optimism should be guarded. Micros are not minis; "computer literacy" is a nebulous term if not measured by a clearly defined standard; and workshops, for all their convenience and popularity, are inadequate for the kind of re-education that library automation calls for. In fact, workshops may be detrimental in leading us to believe that we are becoming computer experts when in reality firm technological control is receding from our grasp. Distinguishing between the microcomputer used as a "toy" and a "tool," Sloan observes,

> End-user programming that's done mainly for the "joy of computing" is often ineffective, inefficient, and a waste of resources. Just because someone spends the better part of the working day hunched over a micro doesn't mean that the end product is an effective system compatible with the organizational goals.[23]

Therein lies the gist of the automation problem. Librarians need access to coordinated educational paths of professional development in the areas of computer skills. The content must be thorough and the distribution extremely broad. The mechanics of this process of re-education are difficult to perceive, but without such re-education, library automation will never reach its full potential. According to Baumgarte,

> Those who learn to use the computer by doing specific tasks and do not understand the overall configuration or functioning of the computer are likely to develop a kind of blind trust or dependency on the computer and computer personnel, resulting in anxiety about the validity of the results and frustration when things do not work as planned.[24]

LIBRARIANSHIP
IN THE LARGER ENVIRONMENT

If increased professionalism is the long-range objective for the proper management of library automation, the short-term strategy should certainly be a carefully controlled exploitation of the expertise currently available only outside the profession. This is perhaps the major finding of the present study. There are many factors and issues to consider and many avenues to achieving the control necessary. The case studies presented here are an attempt to illustrate the fact that the proximity of one's automated resources does not necessarily guarantee their effective utilization. Other issues are involved, other strategies available, other conditions prevalent in the social and political environment of the library that dictate the proper course for acquiring and maintaining control of a highly complex and important resource, the automated system.

The library manager must be sensitive to the socio-political environment and to conditions within the library. Each situation will be somewhat different from all others. And although the trend in library automation is currently to locate the system in house, relying on vendor-trained staff members to operate and maintain a powerful turnkey system, this may not be the best option for every library.

Underlying the dilemmas automation produces for individual libraries are larger issues that affect the library profession as a whole and that, in the end, take a somewhat paradoxical turn. That is, this book has attempted to show that using non-librarians to operate automated systems may be a necessary step toward finally establishing firm computer expertise within the profession.

Librarianship, then, has its own in-house option to consider. This involves (1) an assessment of computer expertise among librarians, (2) an evaluation of automated library systems and the technical expertise required to use them effectively, (3) a recognition of the disparity between levels of competence found, (4) the formulation of plans to alleviate the technological imbalance, provisionally at first, then permanently, and (5) accommodating the consequences of these actions within the framework of the profession.

The principle recommendation of this study is to increase the professionalization of librarianship, specifically by broadening training to include computer literacy, computer science, and systems analysis components, and by intensifying the study of the philosophical principles of librarianship. The findings of the study indicate that library automation and the increased organizational formality it entails normally could lead to professional-organizational conflict, especially if librarianship is attempting to increase its professionalization simultaneously with an increase in automation. In reality, however, the possibility of conflict will be minimized if the reason for the increased formality — automated technology — itself becomes part of the increased professionalization and the diversity of activities that all librarians learn as part of their professional education.

It is clear that the transformation foreseen in the shape and content of the library profession will not be effected overnight. Much time is needed. In the interim, the conditions discussed in this work indicate that libraries should devise methods for enlisting automation expertise from outside the profession, while ensuring complete control of that expertise according to the purposes and services of the individual library.

REFERENCES

1. Ellen Bernstein and John Leach, "Plateau: In Career-Development Attitude Sampling, Librarians See Advancement as Problem," *American Libraries* 16(3) (March 1985):179.

2. Joseph Matthews, quoted in "Automating Libraries: The Major Mistakes Vendors Are Likely To Make," edited by Jon Drabenstott, *Library High Tech* 3(2):112.

3. Bernard G. Sloan, "Micromania: A Manager's Perspective," *Library Journal* 110(12) (July 1985):31.

4. Rob McGee, quoted in "Automating Libraries," p. 110.

5. Michael Stuart Freeman, "'The Simplicity of His Pragmatism': Librarians and Research," *Library Journal* 110(9) (May 15, 1985):28.

6. Michael K. Buckland, *Library Service in Theory and Context* (New York: Pergamon Press, 1983), p. 43.

7. Buckland, p. 136.

8. Buckland, p. 126.

9. Freeman, p. 27.

10. Paul Wasserman, *The Librarian and the Machine* (Detroit, MI: Gale, 1965), p. 30.

11. Richard H. Hall, *Organizations: Structure and Process* (Englewood Cliffs, NJ: Prentice-Hall, 1972), p. 121.

12. Hall, p. 122.

13. Samuel Rothstein, "Why People Really Hate Library Schools," *Library Journal* 110(6) (April 1, 1985):46.

14. Rothstein, p. 48.

15. Roger Baumgarte, "Computer Anxiety and Instruction," *AEDS Monitor* 23(11,12) (May/June 1985):21.

16. Baumgarte, p. 21.

17. Amatai Etzioni, *Modern Organizations* (Englewood Cliffs, NJ: Prentice-Hall, 1962), p. 78.

18. Rothstein, p. 47.

19. Freeman, p. 28.

20. Rothstein, pp. 45-48.

21. Bernstein and Leach, pp. 179-180.

22. Susan Rappaport, "Getting Librarians Involved With Computers," *Library Journal* 110(11) (June 15, 1985):44.

23. Sloan, p. 32.

24. Baumgarte, p. 22.

Bibliography

Ball, Marion J. & Sylvia Charp. *Be A Computer Literate*. Morristown, NJ: Creative Computing Press, 1977.

Baumgarte, Roger. "Computer Anxiety and Instruction." *AEDS Monitor* 23 (11, 12) (May/June 1985):21-22.

Bernstein, Ellen & John Leach. "Plateau: In Career-Development Attitude Sampling, Librarians See Advancement as Problem." *American Libraries* 16(3) (March 1985):178-180.

Bowker Annual of Library and Book Trade Information. 29th ed. New York: Bowker, 1985.

Brod, Craig. *Technostress: The Human Cost of the Computer Revolution*. Reading, MA: Addison-Wesley, 1984.

Buckland, Michael K. *Library Service in Theory and Context*. New York: Pergamon Press.

Capron, H. L. & Brian K. Williams. *Computers and Data Processing*. 2nd ed. Menlo Park, CA: Benjamin/Cummings, 1984.

Carlson, David H. "The Perils of Personals: Microcomputers in Libraries." *Library Journal* 110(2) (February 1, 1985):50-55.

Cheng, Tina T., Barbara Plake & Dorothy Jo Stevens. "A Validation Study of the Computer Literacy Examination: Cognitive Aspect." *AEDS Journal* 18(3) (Spring 1985):139-152.

Clement, Frank J. "Affective Considerations in Computer-Based Education." *Educational Technology* 21(4) (April 1981):28-32.

Corbin, John. *Managing the Library Automation Project*. Phoenix, AZ: Oryx Press, 1985.

Coyle, Mary L. "The Integrated Library in a Timesharing Environment." In *Conference on Integrated Online Library Systems, September 26-27. 1983: Proceedings*, ed. by David C. Genaway, rev. ed. Canfield, OH: Genaway & Associates, 1984.

De Gennaro, Richard. "Integrated Online Library Systems: Per-

_____. "Libraries & Networks in Transition: Problems and Prospects for the 1980's." *Library Journal* 106(10) (May 15, 1981):1045-1049.

Department of Defense Dependent Schools. *Educational Computing: Support Findings and Student Objectives*. DS Manual 2350.1. Alexandria, VA: DoDDS, 1982.

Drabenstott, Jon, ed. "Automating Libraries: The Major Mistakes Vendors Are Likely to Make." *Library High Tech* 3(2):107-113.

Drucker, Peter F. "Managing the Public Service Institution." *College & Research Libraries* 37 (January 1976):4-14.

Epstein, Susan Baerg. "Integrated Systems: Dream vs. Reality." *Library Journal* 109(12) (July 1984):1302-1303.

_____. "Maintenance of Automated Library Systems." *Library Journal* 108(22) (December 15, 1983):2312-2313.

Etzioni, Amitai. *Modern Organizations*. Englewood Cliffs, NJ: Prentice-Hall, 1964.

Freeman, Michael Stuart. " 'The Simplicity of His Pragmatism': Librarians and Research." *Library Journal* 110(9) (May 15, 1985):27-29.

"Freshman Characteristics and Attitudes." *Chronicle of Higher Education* 31(18) (January 15, 1986):35-36.

Gay, Ruth. "The Machine in the Library." *American Scholar* 49 (Winter 79/80):66-77.

Genaway, David C. *Integrated Online Library Systems: Principles, Planning, and Implementation*. White Plains, NY: Knowledge Industries Publications, 1984.

Getz, Malcolm. *Public Libraries: An Economic View*. Baltimore, MD: Johns Hopkins University Press, 1980.

Griswold, Philip A. "Differences Between Education and Business Majors in Their Attitudes About Computers." *AEDS Journal* 18(3) (Spring 1985):131-138.

Hall, Richard H. *Organizations: Structure and Process*. Englewood Cliffs, NJ: Prentice-Hall, 1972.

Hegarty, Kevin. "Myths of Library Automation." *Library Journal* 110(16) (October 1, 1985):43-49.

Howden, Norman. "Programmers and Inverted Files." In *Conference On Integrated Online Library Systems, September 26-*

27, 1983: Proceedings, ed. by David C. Genaway, rev. ed. Canfield, OH: Genaway & Associates, 1984.

Kaske, Neal K. "Studies of Online Catalogs." In *Online Catalogs, Online Reference: Converging Trends*, ed. by Brian Aveney & Brett Butler. Chicago, IL: ALA, 1984.

Luehrmann, Arthur. "Computer Literacy: A National Crisis and a Solution For It." *Byte* 5(7) (July 1980):98-102.

Martin, Lowell. "Emerging Trends in Interlibrary Cooperation." In *Cooperation Between Types of Libraries: The Beginning of a State Plan for Library Service in Illinois*. 16th Allerton Park Institute, 1968, ed. by Cora E. Thomassen. Urbana, IL: Graduate School of Library Science, 1969.

Matthews, Joseph R. "Turnkey Systems: High Risk for Libraries?" *Library Journal* 110(4) (September 1, 1985):133-135.

_____. "Unrelenting Change: The 1984 Automated Library System Marketplace," *Library Journal* 110(6) (April 1, 1985):31-40.

"Obscenities in OCLC Data Cost New York State $11,000." *American Libraries* 12(4) (April 1981):176-177.

Panko, Raymond. "End User Computing and Information Centers." InfoTech Seminars, Management Programs of Hawaii. Honolulu, HI: January 29-30, 1985.

Parker, Thomas F. "Resource Sharing From the Inside Out. Reflections On the Organizational Nature of Library Networks." *Library Resources & Technical Services* 19 (Fall 1975):349-355.

"Public Access Terminals for California Campuses." *Advanced Technology/Libraries* March 1980:6.

Rappaport, Susan. "Getting Librarians Involved With Computers." *Library Journal* 110(11) (June 15, 1985):44-45.

Rawitsch, D. G. "The Concept of Computer Literacy." *MAEDS Journal of Educational Computing* 1978 (2):1-19.

Reynolds, Dennis. *Library Automation: Issues and Applications*. New York: Bowker, 1985.

Rothstein, Samuel. "Why People Really Hate Library Schools." *Library Journal* 110(6) (April 1, 1985):41-48.

Sloan, Bernard G. "Micromania: A Manager's Perspective." *Library Journal* 110(12) (July 1985):30-32.

"Stand-Alone or Shared? Costings and Considerations in Turn-key Configurations." *Library Systems Newsletter* 3(4) (April 1983): 25-27.

Stevens, D. J. "Educators' Perceptions of Computers in Education." *AEDS Journal* 16(1):1-15.

"Universal Lib. Systems, Dataphase, Gaylord Win Circulation Contracts." *Advanced Technology/Libraries* April 1977:1-2.

Veaner, Allen B. "What Hath Technology Wrought?" In *Clinic on Library Applications of Data Processing: Proceedings, 1979: The Role of the Library in an Electronic Society,* ed. by F. Wilfrid Lancaster. Champaign, IL: University of Illinois Graduate School of Library and Information Science, 1980.

Wasserman, Paul. *The Librarian and the Machine.* Detroit, MI: Gale, 1965.

Webb, T. D. *Reorganization in the Public Library.* Phoenix, AZ: Oryx Press, 1985.

Appendixes

DoDDS Instructional Objectives
in Computer Literacy
and Computer Science

COMPUTER LITERACY

Student Will Be Able To . . .

1.0 DEMONSTRATE UNDERSTANDING OF THE CA-
PABILITIES, APPLICATIONS, AND IMPLICA-
TIONS OF COMPUTER TECHNOLOGY

1.1 Interact with a computer and/or other electronic devices.

> 1.1.1 Demonstrate ability to operate a variety of devices
> which are based on electronic logic.
> 1.1.2 Demonstrate ability to use a computer in the inter-
> active mode.
> 1.1.3 Independently select a program from a computer
> resource library.
> 1.1.4 Recognize user errors associated with computer
> utilization.

1.2 Explain the functions and uses of a computer system.

> 1.2.1 Use appropriate vocabulary for communicating
> about computers.
> 1.2.2 Distinguish between interactive mode and batch
> mode computer processing.
> 1.2.3 Identify a computer system's major components
> such as input, memory, processing, and output.
> 1.2.4 Recognize tasks for which computer utilization is
> appropriate.

1.2.5 Describe the major historical developments in computing.

1.3 Utilize systematic processes in problem solving.

1.3.1 Choose a logical sequence of steps needed to perform a task.

1.3.2 Diagram the steps in solving a problem.

1.3.3 Select the appropriate tool and procedure to solve a problem.

1.3.4 Develop systematic procedures to perform useful tasks in areas such as social studies, business, science, and mathematics.

1.3.5 Write simple programs to solve problems using a high-level language such as PILOT, LOGO, and BASIC.

1.4 Appraise the impact of computer technology upon human life.

1.4.1 Identify specific uses of computers in fields such as medicine, law enforcement, industry, business, transportation, government, banking, and space exploration.

1.4.2 Compare computer-related occupations and careers.

1.4.3 Identify social and other non-technical factors which might restrict computer utilization.

1.4.4 Recognize the consequences of computer utilization.

1.4.5 Differentiate between responsible and irresponsible uses of computer technology.

COMPUTER SCIENCE

Student Will Be Able To . . .

2.0 DEMONSTRATE UNDERSTANDINGS OF COMPUTER SYSTEMS INCLUDING SOFTWARE DEVELOPMENT, THE DESIGN AND OPERATION OF HARDWARE, AND THE USE OF COMPUTER SYSTEMS IN SOLVING PROBLEMS.

2.1 Write structured and documented computer software.

  2.1.1 Write well-organized BASIC programs which include the use of color, sound, and graphic statements.
  2.1.2 Write programs which demonstrate advanced programming techniques used to solve problems in business, scientific, or entertainment applications.
  2.1.3 Write programs in an additional high-level language such as PASCAL, COBOL, or FORTRAN.
  2.1.4 Write programs in a low-level language such as machine language or assembler.

2.2 Demonstrate knowledge of the design and operation of computer hardware.

  2.2.1 Demonstrate unassisted operation of at least two different configurations of computers and their peripherals.
  2.2.2 Use a special-purpose computer or computer-interfaced devices to monitor or control events by sensing temperature, light, sound, or other physical phenomena.
  2.2.3 Describe the computer's digital electronic circuitry in terms of binary arithmetic and logical operators.
  2.2.4 Perform vendor-authorized maintenance on the computer system.

2.3 Use computer systems in problem solving.

  2.3.1 Use data processing utilities, including word processing and data base management, in problem solving.
  2.3.2 Translate software from one language to another or to another version of the same language.
  2.3.3 Analyze different solutions to the same problem.

APPENDIX B

COMPUTER LITERACY EXAMINATION: COGNITIVE ASPECT

_____ 1. Which unit below is responsible for displaying the results of a computer program?

(A) Logic Unit
(B) Central Processing Unit
(C) Input Unit
(D) Output Unit

_____ 2. How does a computer solve a problem?

(A) It recalls answers from its memory
(B) It thinks just like a human being
(C) It follows instructions to do what it is programmed to do
(D) It makes decisions based on its knowledge

_____ 3. According to this block of code, how many Xs will be printed?

```
FOR X = 1 TO 100
PRINT "X"
NEXT X
```

(A) 99   (B) 100   (C) 101   (D) 300

_____ 4. What information does a computer require in order to solve a given problem?

(A) The data related to the problem
(B) The instructions given to the computer
(C) The problem and the answer
(D) The data and the instructions

_____ 5. In the following flowchart, which step(s) may be repeated more than once?

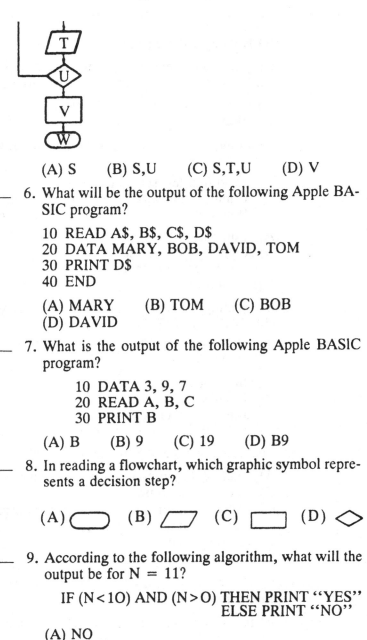

(A) S   (B) S,U   (C) S,T,U   (D) V

_____ 6. What will be the output of the following Apple BASIC program?

10 READ A$, B$, C$, D$
20 DATA MARY, BOB, DAVID, TOM
30 PRINT D$
40 END

(A) MARY   (B) TOM   (C) BOB
(D) DAVID

_____ 7. What is the output of the following Apple BASIC program?

10 DATA 3, 9, 7
20 READ A, B, C
30 PRINT B

(A) B   (B) 9   (C) 19   (D) B9

_____ 8. In reading a flowchart, which graphic symbol represents a decision step?

(A) ⬭   (B) ▱   (C) ▭   (D) ◇

_____ 9. According to the following algorithm, what will the output be for N = 11?

IF (N < 1O) AND (N > O) THEN PRINT "YES"
ELSE PRINT "NO"

(A) NO
(B) YES

(C) N = 11

(D) Nothing will be printed

_____ 10. Approximately, how many bytes of information will be held in a computer with a memory of 10 K?

(A) 100   (B) 1,000   (C) 10,000

(D) 100,000

_____ 11. According to the following algorithm, what will be the final result?

> Read in 5 test scores for student A
> Compute an average score for student A
> Print out the average score for student A

(A) A number

(B) A name and a number

(C) A name and 6 numbers

(D) 5 numbers

_____ 12. In executing the statements below, how many times will HAPPY BIRTHDAY be printed?

> 10 PRINT "HAPPY BIRTHDAY"
> 20 GOTO 10

(A) Only once

(B) 10 times

(C) Infinite number of times

(D) 10 or 20 times depending on the machine

_____ 13. What is a function of program documentation?

(A) To store the program on a disk safely

(B) To ensure program success

(C) To tell the user how the program is organized and what do the variables mean

(D) To provide correct solutions to a specific problem

_____ 14. Which statement below describes a computer program correctly?

(A) It is programmed by a computer operator

(B) It is written in ordinary language

(C) It is loaded into RAM through an output device

(D) It is written to solve a specific problem

_____ 15. Which of the following represents one character in 8-bit microcomputers?

(A) One bit   (B) One byte   (C) 8 bytes   (D) 10 bits

_____ 16. Which is the first step in writing a computer program?

(A) To document the program
(B) To construct an algorithm
(C) To verify program instructions
(D) To write down codes using a programming language

_____ 17. What is the result of the following expression?

6 * 2 + 39 / 3 - 11

(A) 6   (B) 13   (C) 14   (D) 71

_____ 18. Which of the following is a legal representation of computer argument statements?

(A) Z = X
(B) 2A = B - 7
(C) A + B = 3
(D) 2 - 3 = A

_____ 19. Which symbol below identifies a string in BASIC language?

(A) $ (B) #   (C) *   (D) %

_____ 20. According to this Apple BASIC program, what is the final value stored in X?

```
10 X = 400
20 FOR L = 1 TO 3
30 X = X + 2
40 NEXT L
50 PRINT X
60 END
```

(A) 403   (B) 400   (C) 6   (D) 406

_____ 21. What are the major components of a computer system?

(A) Arithmetic unit and input devices
(B) Keyboard and monitor
(C) CRT, language card, and printer
(D) Input device, output device, and CPU

_____ 22. How many bits are contained in each byte?

(A) 8   (B) 10   (C) 16   (D) 20

_____ 23. What is the function of the INPUT UNIT of a computer system?

(A) To process data logically
(B) To read in programs and data
(C) To translate data into machine code
(D) To store data onto disks

_____ 24. Which one of the following is a legal variable name in BASIC language?

(A) 5X   (B) N5   (C) 8A7X6B
(D) 2NDGRADE

_____ 25. What is the output of the following Apple BASIC program?

```
10 FOR L = 1 TO 5
20 READ S$(L)
30 NEXT L
40 PRINT S$(2), S$(1 + 3)
50 DATA ENGLISH, BIOLOGY,
   HISTORY, MATH, SCIENCE
60 END
```

(A) MATH BIOLOGY
(B) BIOLOGY MATH
(C) BIOLOGY ENGLISH HISTORY
(D) ENGLISH BIOLOGY HISTORY MATH
    SCIENCE

_____ 26. What is the last number printed in this Apple BASIC program?

```
10 LET T = 14
20 T = T + 3
30 PRINT T
40 IF T < 22 THEN GOTO 20
50 END
```

(A) 3   (B) 20   (C) 2   (D) 23

_____ 27. Which of the following are binary digits?

(A) 0,1   (B) 0,1,2   (C) 1,2,3,4   (D) 2,4,6,8

_____ 28. Which one of the following is a base two number?

(A) 0110011   (B) 1010102   (C) 123123
(D) 2468

_____ 29. Which of the following accepts instructions and data accessed from the disk?

(A) Permanent Memory
(B) Random Access Memory
(C) Read Only Memory
(D) General Memory

_____ 30. The memory density of a computer is described by K. How many bytes are there in one kilobyte?

(A) 10   (B) 100   (C) 1024   (D) 2000

_____ 31. Which one below represents the correct sequence of computer development?

(A) Vacuum tube, integrated circuit, transistor, microprocessor
(B) Integrated circuit, microprocessor, transistor, vacuum tube
(C) Microprocessor, vacuum tube, transistor, integrated circuit
(D) Vacuum tube, transistor, integrated circuit, microprocessor

_____ 32. Which language will be understood by computers without being compiled or translated?

(A) BASIC language

(B) FORTRAN language
(C) Machine language
(D) COBOL language

ANSWERS

1. D
2. C
3. B
4. D
5. C
6. B
7. B
8. D
9. A
10. C
11. A
12. C
13. C
14. D
15. B
16. B
17. C
18. A
19. A
20. D
21. D
22. A
23. B
24. B
25. B
26. D
27. A
28. A
29. B
30. C
31. D
32. C

APPENDIX C

LIBRARY A:
JOB CLASSIFICATION

Position Title: Computer Technician

NATURE OF WORK

This is skilled technical work involving most phases of operation of computer systems and related peripheral devices. The Computer Technician is responsible for monitoring the operation of the library computer systems, diagnosing problems, and working with maintenance vendors, suppliers of systems in use in the library, and library staff to maintain and improve day-to-day operation and performance. Duties are performed under general supervision of a higher classification. Assignments are evaluated for accuracy and completeness.

ILLUSTRATIVE EXAMPLES OF WORK

- Operates the Central site computer system, such as mounting tapes, setting up printers, backing up disc files, observing console, etc.;
- Receives equipment for repair, clean and dispatch terminals;
- Inventory equipment and purchase necessary supplies;
- Diagnose communication and terminal problems;
- Arrange for repair or replacement of equipment;
- Review equipment vendor manuals to determine operating principles of computer and related peripheral devices;
- Keep current the Computer Operator manuals, reflecting changes in operating requirements and techniques;
- Train other library staff in operation and use of the computer system and related peripheral devices;
- Perform related work as required, such as data entry, etc.;
- Schedule may include evenings, weekends, and on-call work.

KNOWLEDGE, SKILLS AND ABILITIES

- Considerable knowledge of the practical application of computer operation and procedures;
- Working knowledge of data processing procedures and terminology;
- Ability to comprehend and apply the information found in technical manuals and related literature with judgment and initiative to troubleshoot equipment and communication problems;
- Planning and scheduling skills;
- Ability and skill to establish and maintain effective communication and working relationships with other employees and system users;
- Ability and skill in training and communication skills;
- Ability and skill to organize work and carry through established procedures and work plans to meet scheduled deadlines;
- Data entry skills;
- Programming skills in some high-level language.

MINIMUM EXPERIENCE AND TRAINING

Must be a high school graduate or equivalent; two (2) years experience in computer operation, preferably supplemented by vocational training in data processing, with some emphasis on basic hardware maintenance and troubleshooting; or any equivalent combination of experience and training. Experience and training must include responsible work operating a combination of computer and auxiliary equipment, in equipment in an operating environment comparable to that of this library.

LIBRARY B:
COMPUTER OPERATOR III

Description of Work

General Statement of Duties: Performs skilled work in the operation of electronic computers peripheral and auxiliary equipment.

Supervision Received: Works under the supervision of a technical supervisor.

Supervision Exercised: Exercises supervision over assigned personnel incidental to the other duties.

Examples of Duties. (Any one position may not include all of the duties listed nor do the listed examples include all tasks which may be found in positions of this class.)

- Independently sets up and processes routine applications or programs on the computer; assists other personnel in operations involving complex computer processing.
- Prepares the computer for operation by use of computer console and by readying of input and output devices; starts and monitors ongoing operations.
- Interprets and independently responds to well-defined instructions, operating procedures and programs which involve a sequence of operations, exceptions, and controls requiring judgment, initiative, and alertness to assure proper machine functions.
- Performs a variety of tasks on data processing equipment including teleprocessing and other remote devices selecting appropriate data to be used; operates various types of peripheral data processing machines; independently operates small computers in a satellite installation.
- Sets up jobs from library files, audits finished runs for accuracy and control, and performs similar data control or library work related to computer operations.
- Assists in determining machine scheduling requirements; assists auxiliary machine operators; maintains machine usage logs and records for data processing production and related work; assists in shift supervision at a satellite or remote installation.
- Performs related work as required.

Qualifications for Appointment

Knowledges, Skills and Abilities: Working knowledge of the operations, care and adjustment of electronic computers, peripheral and auxiliary computer equipment. Working knowledge of data processing procedures and terminology. Ability to reason and think logically. Ability to apply judgment and initiative to solve problems arising in computer operations. Ability to follow oral and written instructions and procedures. Ability to establish and maintain effective working relationships with other employees and City agencies.

Education: No formal education required.

Experience: Three years of experience in general clerical and computer operations work, at least two of which have been progressively responsible work operating a combination of electronic computer and auxiliary equipment including equipment in an operating system comparable to that in the Data Services Division.

Sixty semester hours from an accredited college or university, at least nine of which were in the computer sciences, may be substituted for six months of the experience requirement. The completion of an 180-hour course in data processing from a technical or vocational school may be substituted for six months of the experience requirement.

LIBRARY C:
LIBRARY MINI-COMPUTER OPERATOR

Nature of Work

The Library Mini-Computer Operator must have a complete and thorough knowledge of Library procedures, as well as the technical skills to operate a mini-computer. Work entails entering data on books and patrons, plus the operator must make additions, deletions and changes to these systems. Operator must be able to load and unload cards and labels on the printer. The operator must be a "trouble shooter" in case of equipment error and interpret indicator error lights and correct the cause of such errors.

Examples of Work Performed

- Enters data into the Mini-Computer on all books for the Library System.

- Enters data into the Mini-Computer on all patrons who register with the Library System.
- Enters data into the Mini-Computer on all additions, deletions and changes for books and patrons in the System.
- Operate a cassette tape device, as well as the magnetic tape device.
- Load and unload cards and labels on the printer.
- Produce catalog cards and labels for all library books in the System.
- Performs other duties as required.

Necessary Knowledge, Skills and Abilities

- A complete and thorough knowledge of Library Procedures.
- Technical skills to operate a mini-computer.
- Ability to insure that the data is correct before sending to Data Processing Department.
- Ability to maintain a small library of cassette tapes and magnetic tapes used by the computer.
- A working knowledge of the basics of data communication devices which require patience and accuracy.
- Ability to line up the print characters on a card or label in order to present a neat, professional appearance.
- Ability to be a "trouble shooter" in case of equipment error.
- Ability to maintain an inventory of cards, labels, sufficient for incoming books.
- Ability to train another employee as a back-up for the Mini-Computer.

Desirable Training and Experience

A high school education, preferably with some business training or experience. Accurate typing skills.

LIBRARY D:
LIBRARY COMPUTER SPECIALIST

General Responsibilities

Under general supervision, to oversee the daily operation of the Library's computer center, including operating the computer; to write, test and debug programs; and to do related work as assigned.

Typical Duties

1. Oversees the daily operation of the Library's computer including scheduling work to be done by the section; sets up computer each day and runs some special programs; determines if computer problems are caused by programming errors, improper operating procedures or machine malfunction and takes action to correct or arranges for repair services; maintains computer and peripheral equipment, arranges for needed servicing or reports major equipment problems to supervisor; plans, assigns and reviews work of others in the center, trains new employees and assists Department Head with other supervisory functions.

2. Assists Department Head in determining when new jobs can be done by the center and most efficient way of doing them; prepares flow charts for new work; writes, tests and de-bugs new programs in the language required by the Library's computer; makes modifications to existing programs, recommending significant changes to improve operations, and executing approved changes as assigned; assists Department Head in maintaining documentation for each job and/or computer application; may participate in research of new systems applications; maintains adequate inventory of supplies for the center and requisitions for replacements; performs other duties as assigned.

Minimum Requirements

1. Training and Knowledge Required for Job Application: Considerable knowledge of computer operation and maintenance methods, procedures and techniques; considerable knowledge of procedures and methods used to prepare and revise computer

programs in language required by the department's computer (for example, RPG2); some ability to assist in developing new computer applications and research new systems applications; considerable knowledge of departmental policies and procedures.

2. Knowledge and Ability Necessary for Full Job Performance: Thorough knowledge of computer operation and programming as defined above; ability to assist with systems analysis and related programming applications specific to the department; ability to act as lead worker and schedule, oversee, review and train other employees in the work of the section; ability to maintain records and prepare reports on staff and machine performance and production; ability to analyze problems, and take necessary corrective actions; ability to develop and maintain effective working relationships.

LIBRARY E

POSITION TITLE PROGRAMMER ANALYST	DEPARTMENT DATA PROCESSING
	DIVISION DATA PROCESSING
TITLE OF IMMEDIATE SUPERVISOR DIRECTOR OF DATA PROCESSING	

1. Purpose of Work

Under the general supervision of the Director of Data Processing, performs systems analysis, programming, program documentation and other technical tasks related to data processing and is responsible for the overall operations of the department during the absence of the Department Director.

2. Minimum Qualifications

1. Associate degree in Data Processing or equivalent combination of education and experience.

2. Demonstrated ability to analyze, organize and think logically.
3. Two or more years experience in a responsible RPG II programming position.
4. Knowledge of data base concepts, structured programming and other languages.
5. Good communication skills.
6. Demonstrated supervisory and organizational ability with respect to data processing.
7. Possession of a valid driver's license and ready access to suitable transportation or a personal vehicle appropriate for use in the transaction of Library business and performance of job related duties (approved mileage reimbursed at prescribed rate).

3. Typical Duties

1. Works with the Director of Data Processing and library staff in the development and implementation of computer systems.
2. Responsible for overall computer programming, program testing and program maintenance.
3. Responsible for overall systems, programming, and operations documentation.
4. Assists or substitutes for the Director of Data Processing; develops and monitors work of designated personnel in the Data Processing Department.
5. Performs other duties as assigned.

POSITION TITLE COMPUTER OPERATOR/DATA ENTRY OPER.	DEPARTMENT DATA PROCESSING
	DIVISION DATA PROCESSING
TITLE OF IMMEDIATE SUPERVISOR DIRECTOR OF DATA PROCESSING	

1. Purpose of Work

Under the general supervision of the Director of Data Processing, operates and monitors computer; operates unit record and peripheral equipment; performs clerical and technical tasks related to processing of data. Operates alphabetic and numeric keypunch machine to transcribe data from source material onto punch cards and produce prepunched data.

2. Minimum Qualifications

1. High school diploma, G.E.D., or equivalent experience.
2. One year of experience with tabulating equipment or completion of a recognized operations instruction course in electronic computer system.
3. Ability to deal with problems involving several variables in familiar context.
4. Demonstrated ability to analyze, organize, and think logically.

3. Typical Duties

1. Operates and monitors computer on pre-scheduled production runs.
2. Performs clerical and technical tasks related to processing of data and operating computer and peripheral equipment (e.g., maintains record of runs and malfunctions; distributes output, etc.)

3. Operates unit record or tabulating equipment; operates other computer-related equipment.
4. Keys and verifies data from input documents according to job instruction sheet.
5. Performs other miscellaneous duties as assigned.

POSITION TITLE COMPUTER SYSTEMS LIBRARIAN	DEPARTMENT DATA PROCESSING
	DIVISION DATA PROCESSING
TITLE OF IMMEDIATE SUPERVISOR DIRECTOR OF DATA PROCESSING	

1. Purpose of Work

Under the general supervision of the Director of Data Processing analyzes library procedures and develops new information systems to meet current and project needs.

2. Minimum Qualifications

1. A Bachelors Degree and a Masters Degree from an ALA accredited college or university.
2. General knowledge of and appreciation for computers and automation.
3. Ability to prepare comprehensive reports and present facts concisely.
4. Ability to communicate effectively orally and in writing.
5. AT LEAST TWO (2) YEARS OF EXPERIENCE IN A PUBLIC LIBRARY PUBLIC SERVICE POSITION, PREFERABLY IN A SUPERVISORY POSITION.
6. Possession of a valid driver's license and ready access to suitable transportation or a personal vehicle appropriate for use in the transaction of Library business and performance of job related duties (approved mileage reimbursed at prescribed rate).

3. Typical Duties

1. Plans and prepares technical reports, memoranda and instructional manuals, relative to the establishment and functioning of complete operation systems.
2. Plans and helps design detailed information systems in cooperation with appropriate departmental personnel.
3. Studies existing data handling systems to evaluate effectiveness and develop possible alternate.
4. Develops cost analyses of new or old systems to assist in determining the feasibility and extent of library automation.
5. Collaborates with and acts as a liaison between the library and providers of contract automated services.
6. Performs other duties as assigned.

POSITION TITLE DIRECTOR OF DATA PROCESSING	DEPARTMENT DATA PROCESSING
	DIVISION DATA PROCESSING
TITLE OF IMMEDIATE SUPERVISOR EXECUTIVE DIRECTOR OF THE LIBRARY	

1. Purpose of Work

Under the general supervision of the Executive Director, provides direct liaison to user within the library system; schedules and coordinates processing of data; controls accuracy and flow of data and provides supervision to staff.

2. Minimum Qualifications

1. BA or BS in business administration or related area or the completion of four years of electronic data processing technical school or other equivalent combination of training and experience.
2. Three to five years of related technical experience involving

the processing and programming of data through a computer system.
3. Demonstrated managerial skills (supervision, training and development).
4. Possession of a valid driver's license and ready access to suitable transportation or a personal vehicle appropriate for use in the transaction of library business and performance of job related duties (approved mileage reimbursed at prescribed rate).

3. Typical Duties

1. Develops, maintains and controls data processing to insure timely and accurate completion of jobs, provides assistance and information to users.
2. Supervises data processing unit; plans and schedules work; establishes quotas, assigns work, maintains production and work flow, directs, reviews, and evaluates work.
3. Analyzes data, debugs programs and makes program modifications.
4. Is responsible for preparation of administrative reports, budgetary requests, and selection evaluation and training of assigned personnel.
5. Performs other duties as assigned.

LIBRARY F:
CLASSIFICATION SPECIFICATION

Classification: Systems Analyst 15

Distinguishing Characteristics: This is supervisory or advanced level computer oriented systems analysis and design work. Employee in this classification works on complex internal management information problems involving all phases of systems analysis. Problems are complex because of diverse sources of input data and multiple-use requirements of output data in which every item of each type is automatically processed through the full system and appropriate follow-up actions are initiated by the computer. Employee may make recommendations for approval of system installations or changes. Employee provides functional direction as a project leader or coordinator to lower level Systems Analysts assigned to assist. Assignments are received and work is reviewed by a higher level supervisor or department management personnel for conformance with city policy and departmental objectives.

Typical Duties:

1. Performs administrative duties to develop, recommend action and implement assigned portions of or complete operating systems, interfunctional systems within a program or organization, or systems which integrate the operations within a functional area.
2. Plans analysis activities as a project leader and conducts comprehensive investigations to determine the need or to evaluate internal departmental requests for new or modified systems; determines nature and depth of analysis required for additional substantiating studies.
3. Determines operating problems, existing work flow and operational relationships within a functional area or between several related functional activities.
4. Consults with appropriate supervisory personnel of organizational units to decide between possible courses of action pertaining to systems development.
5. Develops new or improved operating systems and methods as part of a full systems concept or as a full functional systems plan; analyzes basic policy requirements; outlines the scope, objectives, work flow and responsibility assignments of personnel; and anticipates possible conflicts within the system being developed or between functional activities.
6. Adapts the development of systems to accommodate application of computer equipment and coordinates interface and compatibility problems with cognizant programming personnel; coordinates the development of test problems and participates in trial runs of new or revised systems.
7. Plans, prepares, and makes presentations to management of proposed new or revised systems and operating procedures;

proposes alternate solutions to unresolved problems, methods of implementation, and necessary changes to department or city policy; conducts economic feasibility studies of proposed changes and submits reports; obtains acceptance and agreement as to application and installation of proposals.

8. Coordinates the implementation of approved systems, as project leader, to ensure a smooth transition and installation of new or improved systems by training others, developing supplementary information and by collaborating with concerned personnel to resolve questions in such areas as application or operational difficulties.

9. Performs related duties as assigned.

MINIMUM EXAMINING REQUIREMENTS

Education: Bachelor's degree in business, economics, mathematics, computer science, management science, statistics, engineering, or a closely related major field or equivalent.
Experience: Five years progressive systems analysis and design experience.

Acceptable Equivalency:

Abilities Required to Perform Work: Thorough knowledge of systems analysis methods and techniques; thorough knowledge of digital computer capacities characteristics and design features; ability to deal analytically and systematically with problems of organization, workflow, analysis of information requirements, and planning of integrated procedural systems; knowledge of practices of larger organizations; ability to relate all elements of complex problems into systematic solutions; ability to develop new systems methods and techniques; leadership; ability to train others in systems design; ability to plan and schedule work activities of other analyst assigned to a project; ability to relate to people in a manner to win confidence and establish rapport; persuasiveness in order to gain the cooperation of others and their acceptance of proposals regarding the implementation of new or revised systems; flexibility to adjust to changing conditions with ability to make decisions; verbal facility and ability to relate to people in consultation or gathering of information.

Working Conditions and Hazards: Normal office conditions.

Special Requirements:

Promotes from: Systems Analyst 13
Promotes to: Systems Analysis Manager 17

Approved: Date:

Index

Printed in the United States
by Baker & Taylor Publisher Services